ENERGY AND EXERGY CURRENTS

(An Introduction to Exergonomics)

E.I. Yantovskii

ENERGY AND EXERGY CURRENTS

(An Introduction to Exergonomics)

NOVA Science Publishers
1994

Art Director: Christopher Concannon
Graphics: Elenor Kallberg and Maria Ester Hawrys
Book Production: Roseann Pena, Tammy Sauter, and Michelle Lalo
Circulation: Irene Kwartiroff, Annette Hellinger,

Library of Congress Cataloging-in-Publication Data

Yantovskii, E.I.
Energy and exergy currents : an introduction to exergonomics / E.I. Yantovskii.
p. cm.
Includes bibliographical references and index.
ISBN 1-56072-175-8 : $69.00
1. Power (Mechanics). 2. Thermodynamics. I. Title.
TJ163.9.Y36 1994 94-6545
621.042--dc20 CIP

6080 Jericho Turnpike, Suite 207
Commack, New York 11725
Tele. 516-499-3103 Fax 516-499-3146
E Mail Novasci1@aol.com

Printed in the United States of America

To my dear Ilia, Nina and Serge.

CONTENTS

PREFACE

This book was written in Russia. This is a collection of recently published papers, most of which have been delivered at International Conferences (e.g., the World Clean Energy Conference in Geneva, 1991, the International Conference on CO_2 Removal in Amsterdam, 1992, ATHENS '91, ECOS '92, ENSEC '93). Some of the material has served as lecture topics in the U.S.A., Sweden, the Netherlands, Norway and China.

The word "exergonomics" I'm responsible for might be considered by somebody as a language misuse. It reflects my attempt to underline a similarity to ordinary economics, but to stress the principal difference due to the lack of monetary values replaced by exergy. "Without a constant misuse of language there cannot be any discovery, any progress " (Kelvin). As I hope, in some future energy policy of states and companies will be closely related to exergonomics, especially of renewable energy sources and clean fuel-fired power plants.

I completed work on this manuscript in the Netherlands, where I had the good fortune to spend my vacations.

The scenery of this wonderful country is inspiring. With its perfect universities, industry, magnificent agriculture and high standard of living, it presents an invaluably positive example for the whole world.

In tribute to the Netherlands, I offer the words of a great Dutchman, Erasmus, which may serve to shed light on the contemporary problems of authors impudent enough to write books in a language other than their mother tongue:

...now grammar alone is quite enough to convert all human life to a continuous state of torture.

XXX II.

Theologians experience hunger, physicists catch cold, astrologists feel ashamed... The medical man alone is ahead of the others.

XXX III.

...how scornful they look at the laymen, installing triangles, circles, squares and other mathematical figures to raise a dust on profanes.

LII.

(Des. Erasmi Roterodami, Stultitiae Laus, Anno 1511).

My thanks to the fine work of the editors of NOVA Publishers. My deviations from the standard rules of English grammar are not always evident to me. So I escaped a torture of grammar.

Also, thanks to the warm hospitality of the family that sheltered me in Holland, I managed to stay healthy and thus eluded the good intentions of the medical profession.

I did my best to be explicit, not to raise a dust on anybody.

Finally, I would like to express my gratitude for the patience, understanding and spiritual support of my wife, Alla, who accompanied me to Holland. Only by abbreviating our enjoyment in seeing Holland's sights, could I finish my work on this book.

Haarlem–Moscow–Haarlem
27.02.1992 26.07.1993

NOMENCLATURE

A = magnetic potential
B = magnetic induction
D = exergy losses, electrical displacement
E = electric field
ρe = energy per unit volume
F = conductor cross-section
G = chemical potential
G_e = chemical exergy
H = magnetic field
J = electric current
j = current density
K_e = net-exergy coefficient
L = length
M_{ik} = electromagnetic tensor
P_{ik} = mechanical tensor
P = pressure
Qu = share of heat
q = space charge
R = electrical resistance
S = entropy
T = absolute temperature
U = internal energy
V = voltage, velocity, volume
W = work share
Z = energy intensity, SUSEX criterion
δ = energy current density
δ_ε = exergy current density
ε = exergy
λ = thermal conductivity
$\eta_2=\eta_\varepsilon$ = exergy efficiency
σ = electrical conductivity
σ_s = entropy gain intensity
τ_{ik}=tangential stress
φ = electric potential
ρ = mass density
∇ = nabla-operator, $\nabla\cdot(\ )$=div ( ), $\nabla x(\ )$=rot( ); $\nabla(\ )$=grad ( )
Δ = finit difference symbol

Subscripts
q = electrical charge
s = entropy
i, k = tensor symbols

For a correct solution to social problems it is not sufficient to be familiar with political doctrines; the skill to properly read the book of nature is necessary.

It's impossible to recycle totally dissipated energy. This means that energy is degraded in natural processes. The reason for the irrevesibility of natural processes and for the degradation of energy is the change from large and intensive motions to small slower ones, gradually disappearing into the small, chaotic, orderless motion of material particles.

The final result of all processes in inorganic matter is an increase in disorder, referred to as a gain in entropy.

N.A. Oumov, 1906

Chapter 1.

OUMOV'S THEOREM

What is poetry? I don't know. But if I knew what it is, I would have never been able to express my knowledge, or finally if I could to choose and bring together adequate words in any case, I would have never been understood by anybody. There exist such the realities which it had better not to specify at all.

Innokenty Annensky

In my life I've never seen such abracadabra.

From official opinion to doctoral thesis of N.A. Oumov's "Equation of Energy Motion in Bodies" 1874.

Vector of the current density of energy

What is energy? I don't know. Sometimes it is defined as a measure of intensity, which adheres strictly to a law of conservation. It is possible that energy, like poetry, is a reality which is better left undefined.

One need not worry if a term can't be expressed by means of other, clearer terms. There are many such terms in physics (time, space, electricity, etc.), which can't be expressed using something simpler. These are merely general concepts and we become accustomed to their varied usage. These terms are connected by the Laws of Nature which allow us to predict phenomena thereby allowing us to design machines to use these principles to our advantage.

Energy can only be measured by quantitative means. We often speak of kinetic, chemical or nuclear energy therefore we must bear in mind that, in effect, these are only energy carriers: the motion of the body as a whole, the strength of bounds of atoms in molecules or particles in nucleus.

The very important internal energy U is the sum of kinetic energy of haotically moving particles. Sometimes internal energy is mistaken for heat, which we will discuss later.

However, in ancient times, the theory of "caloric", a substance which enters the body when it is heated, was very popular. It has penetrated so deep in the human mind, that in Russia there has been some accidents, which is described by the eminent Russian scholar, M. Lomonossov, more than two centuries ago. His words are a good illustration to the internal energy term:

> Some priests, not only in villages but also in towns, used to make the Christian bathe of babies in the very cold water in winter, sometimes with ice, quoting the manual requirement to use clean, natural water without contaminations. They treat heat as the contaminating substance, forgetting that in the summer they do the same in rather warm water, dirty as they think. Being ignoramuses, they aren't aware of the amount of heat, which really exists in very cold water.
>
> Such obstinate priests which forced to bath the babies in ice water I consider as the executioneers, who want after the birth soon to be invited to funerals for their good.
>
> M. Lomonossov, "On the Conservation and Reproduction of Russian People" 1761.

In 1874, N. Oumov had proposed to describe the energy flows by means of a vector in energy conservation law. In 1884, J. Pointing had wrote the energy conservation law in electromagnetic processes, using famous Maxwell equations, where the energy current vector for field has been introduced.

For us, the Oumov-Pointing vector of the current density of energy plays a major role. We shall consider not only the vector as a whole, but also its components, which describe the motion of different energy carriers. At first we will use a rather simplified form of the vector and restrict ourselves the energy flow in the substance only, assuming all the processes as steady-state. Then the more general case will be mentioned.

Let us take as starting point the following easy remembered form:

$$\delta=\varphi j_q+V\cdot P+G\rho V+Tj_s \tag{1.1}$$

The dimension unit for this vector is the relation of power to surface, in SI system it is W/sq.m.

The symbols in this equation mean:

φ = electrical potential. Its difference in two points means electrical voltage: it is well known and easily measured in practice.

V = velocity of motion. Here it plays a role of potential, in $V\cdot P$ term.

G = chemical potential, referred to as Gibbs energy also. It represents the strength of bounds of atoms in molecules. Its values are sampled in chemical handbooks for large amounts of species.

T = thermal potential or absolute (Kelvin) temperature.

As a rule, all of these potentials are scalars; they can't show the direction in space. But for velocity, sometimes, the exemption is needed, as in the case of energy transfer by impulse in solids or viscous liquids. We will see this later, considering the full system of equations. All of the terms, defined as j with a subscript, are the energy carriers current density:

j_q = electrical current density. It is uniformly distributed over the cross-section of a conductor F and the full current $I=j_qF$. This current is the flow of electrical charges, for example electron gas flow relative to standing ions in models of metal conductors.

P = impulse flow density. This describes the intensity of motion and obeys strict conservation law, the second Newton law. The product $V \cdot P$ is mechanical current (power) density of energy. Speaking of the ability to do work, we have to keep in mind the ability to transfer energy of other carriers to impulse flow. Energy transfers from motor to machine-tool or from the turbine to generator via the shaft by impulse flow. Very often the direction of velocity differs that of energy. Here we need to consider the impulse as a tensor.

The most important concept is the thermal charge current density or entropy current. In effect entropy is multiplied by the constant logarithm of a number, which reflects how many different positions have the considered physical system. But for the sake of symmetrical description, we'll consider it as the same among other energy carriers like electric charge, substance and impulse. If entropy flow is uniform, it equals $\dot{S} = j_S F$, and the total heat flow is $Q = \dot{S} T$.

Oumov was first, who introduced energy vector, like δ and wrote the energy conservation law in the form of

$$\frac{\partial \rho e}{\partial t} + div\delta = 0 \tag{1.2}$$

This is the special form of differential equation (divergent form) which reflects strict conservation law of any quantity if the right hand side (rhs) is zero. Otherwise, if there exists the gain (birth) or the sink of the quantity, the rhs is positive or negative. It is just Oumov theorem.

When we used to remember the great achievements of R. Mayer, H. Helmholz and J. Joule as the founders of energy conservation law we have to keep in mind that in vigorous form of differential equation the law was expressed by N. Oumov in 1874, when he was 28.

Strictly speaking the δ in (1.1) is valid for the steady-state processes, where $\frac{\partial}{\partial t} = 0$, nothing changes with the time. In this case we have $div\delta = 0$. Such vectors referred to as solenoidal ones, their vector lines are going from infinity to infinity, they can not brake off, they can form closed loops only.

Oumov's Forecast

"Political ability is the ability to foretell what is going to happen...next month and next year. And to have the ability afterwards to explain why it didn't happen."

Winston Churchill

N. Oumov has the outstanding ability to look into the future to find the proper ways for human to develop clear in his article, written in 1912. Let us quote it.

> The power needed to mankind equals 2300 billions of HP.h, from which 20% is spent to mechanical force, 40% for heating and 40% for metal production and industry, minor part for lighting. This distribution differs from divisions of coal use. Namely of 1250 mt of coal in 1910 nearly 50% was used to produce the mechanical forces, 30% for heating, 20% for metallurgy and industry, nearly 3% for light.
>
> In the force production the waste of fuel (extra fuel) is 600 mt of coal, in heating 270 mt, in lighting 35 mt. All in all the 1050 mt is extra spending, nearly 80% of total input... It leads to the necessity to increase the efficiency of heat engines.
>
> The hydropower of the world is guessed as 1-2 billions of HP.h, three times more than mechanical energy demand, but it equals only one half of the total energy consumed on the earth. Therefore, this source cannot maintain and develop modern culture.
>
> We have to think about new ways to create energy. We can't count on irregular power, like wind or ocean waves. Nevertheless the latter develop in storms hundreds of thousands HP per kilometer of the coastal line. The tide seems to be better. The tidal wave reaches, in some places, a height of 5-10 m. Perhaps, it would be wise to build large water storages, which are filled when the tides are high, discharged when the tide is off, driving hydraulic turbines. The tide power in effect is the use of rotational kinetic energy of the earth. This use slows the rotation of the earth and creates a day elongation. However, the amount of kinetic energy is so great, that even if it was increased one hundred times more than contemporary global, the day would increase one second every ten thousand years. There exists no harm in the use of kinetic energy of the earth for mankind.
>
> All of these energy sources on earth, including internal heat, transferred to surface and irradiated into space are insufficient or very difficult to use.

We have to look at the heaven, where we find our Sun, giving us a tremendous flow of energy and creating all of our energy resources. In the coal, hydropower, wind and waves we deal with the stored energy of sun beams. Hitherto we use this energy in an unwise manner. It leads us to our doubts. However it remains, if we put the question on direct use of solar beam energy. There is nearly 1700 kcal/sq. m.h, falling. This energy flow decreases in the air, being absorbed mainly by water vapor and carbon dioxide, water droplets, clouds, dust, etc. Absorption increases when we move toward the earth poles, when the angle of beam inclination to the earth surface decreases.

At the latitude 45°, the power equals nearly one HP/sq.m at least in summer, 9 hours daily. If we assume this figure in the Sahara desert, the result will be less than actual: we find that in 1/3 of daytime, the Sun brings to Sahara 9 billions* of HP, or 3-4 billions if uniformly distributed over the day. If we assumed that only thousandth part is consumed, we would have 30 billion HP.h annually. It is ten times more than contemporary energy consumption, involving heating as well.

Therefore the Sun is that energy source where, in the far future, mankind can and will get the energy needed for its cultural being and development. Hitherto there exists only a small attempt to direct use of solar energy. In California, USA, and some other states, the concentration of solar beams by means of concave mirrors on a surface of small boilers of engines, using steam or ether vapor is being investigated.

However, these devices are out of the main way of expected use of solar radiation. The temperature of the sun exceeds 5000°C, the temperature difference between the Sun and Earth is so great, that solar energy is almost all might be converted into mechanical power. This property of solar radiation is ignored by the approaches which mean all the task in boiler heating only.

It is possible to say that mankind will use solar energy in quite another manner, by means of electrical processes. Still, everything has to be done in this problem. It will take some centuries of time before we reach a solution, however.

The misfortunes and failures are waiting for us, but our outstanding goal is worth the efforts and victims, because the future of mankind depends upon its achievement [p. 449-461].

---

* Here, in Oumov text, billion=10^{12} (E.I.Y.)

After this long citation we can wonder how contemporary are the words of N. Oumov, written eighty years ago. Here we can see the importance of figures of the current density of energy and clearly underline the large temperature difference (between the Sun and the earth) as the condition of convertibility of energy. In this book we'll try to consider these topics in some detail.

Here, we have to point out that California is still a leading region in solar energy use. The first commercial solar power plants assisted by natural gas was put in operation on December 28, 1989. It is developed by Luz Company and consists of some 80MW units with concave mirror solar concentrators, heating special liquid (difenil) in a pipe.

Certainly, the part of Oumov's figures are now changed. The world annual energy consumption was increased from 1,25 btce to 10 btce at present. But quantitative change can't undermine Oumov's ideas, such as electrical processes of direct solar energy conversion. It is just photovoltaics, as we refer to it now. While looking at modern books of J.O'M. Bockris (1980) or many other advocating solar-hydrogen options, based on the photovoltaics progress, we have to remember Oumov, too.

The electrical processes, mentioned by Oumov, photovoltaics are widely used now. Annual production of solar photovoltaic cells reaches tens of megawatt. The sad joke of Mr. Churchill doesn't relate to Oumov's foretell, in spite of the greater time span of forecasting, not months or years, but centuries, in this case a man can't be an eyewitness of the victory of his mind.

Some Examples

W.Ebeling has presented a clear image of the flow of energy in the Universe: energy flow appears in thermonuclear processes inside the stars. It is transferred by means of radiation (photon currents) to all the space bodies, including planets. Ebeling refers to this flow as "photon rain". The planets, having lower temperature and increased entropy, reflect radiation and eventually it equals the relict, residual radiation of temperature 3-10 K only. Photon rain can only fall down, it is the direct consequence of the Second Law of Thermo dynamics.

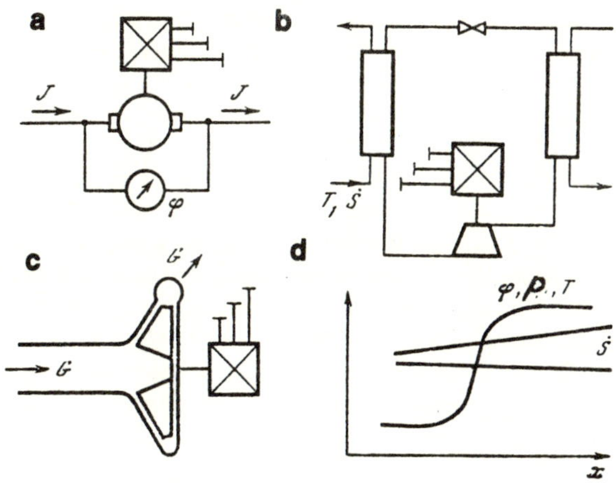

Figure 1.1. The potentials and flows in some devices.
a) Direct current generator
b) Hyraulic pump
c) Heat pump
d) Potentials and currents along a path.

We can see the gradual or stepwise change in energy currents in all the engineering as well as in nature. The main energy sources at present, as it was in Oumov times, are substances of high chemical potential, fuels, which need to be mined and delivered to conversion devices. The energy rider changes the horses, e.g. from fuel energy current GρV to heat current Tj_S in boiler, then to impulse flow in turbine and shaft to generator.

Let us consider three engineering schemes (Fig. 1.1) with changing of energy currents: a) electrical generator, direct current, b) water pump, c) compressional heat pump.

In the a) case, the electric charge flow is going to the generator from the left, then, inside the generator, the charges are moving against electric field, where driving power is consumed. In any electrical generator an external (electromotive force) influence is needed to drive the charges against the electric field. Moving charges enter the regions of higher electric potential, with which they leave the generator and go right. The total electric current from left to right remains the same, however the potential in the right side much exceeds the left. We used to say that the generator delivers the electrical energy to the line.

In the b) case, the flow of water is coming from the left, causing the pressure inside flow to increase thus creating the impulse flow to rise which is transferred to water by the blades of rotating impeller of the pump. The energy flow increases along with the impulse flow whenever the flowrate of water is the same on left and right. Energy flow on the right side is much higher. We used to say that the pump delivers hydraulic power. Here, the potential V is the same and power increases on account of impulse flow, the force action of blades.

In the c) scheme, heat pump, some elements are working together in a closed loops, filled by easy boiling liquids, like freons.

There exists a lot of freons, and their boiling points differ in a wide range. The elements of the device are: two heat exchangers (evaporator left and condenser right), a compressor with driving motor, and the throttle, hydraulic resistance. Here, the low grade heat flow is going from the left to the evaporator, where it is absorbed by the boiling freon of low temperature T_1. The vapor is compressed, its temperature increases to T_2, then the vapor condenses to liquid. The condensation heat is transferred to water flow in a heating system. The vapor-liquid mixture of low pressure enters the evaporator and everything repeats.

The entropy flow here is coming from the left with low potential (say T_1=280 K) and leaving with an increased one (T_2=350 K), therefore the heat gets new property; it can heat the house. Meanwhile, no heat flows of low grade however large they are, could do the same because of low temperature. In the heat pump appears the similarity of the temperature difference to electric voltage, in other words, the analogy between the thermal and electrical potential. Sometimes, the heat pump is referred to as a heat transformer in view of that analogy.

Let us depict (Fig. 1.1 d) the evolution of above mentioned devices in explicit graphs. Let X be the coordinate of our ordinary space along energy carrier current. The vertical axis serves as potentials. In the a) case the potential of the flow is risen; in the b) case the carrier flow increases, the impulse flow, and in c) case, both potential T and energy carrier entropy current j_s are raised.

Each of the three might be considered as a thermodynamic system. The closed boundary surface is assumed around it. The currents across from the boundary have to be calculated. The system is a steady-state, no one quantity depends upon time, energy can not be stored, therefore inflow and outflow of energy are the same.

In all three cases, the two flows are at the entrance: energy flow from the left and an electric energy flow driving the motor of the electric generator, water pump or compressor of the heat pump. At the exit, also, two flows exist. The aimed flow, which is the reason of the device and the heat flow, caused by entropy gain. Therefore it's possible to speak about the same structure of energy flows in all three examples.

The heat which appeared is a negative phenomenon. It usually is referred to as "energy losses". Certainly, energy losses are impossible, according to First Law. Here, unavoidable entropy gain leads to conversion of a part of the energy flow into heat. The sum remains constant, whereas the aimed flow of energy is decreasing. This undesirable heat might be small or large, therefore, desirable energy flow at the exit of the device is large or small. How much might it be? In particular, how much work could we get from the entropy flow?

We'll try to answer these questions in the next chapter.

Chapter 2.

EXERGY

I believe in God and Gibbs.
A. Einstein

Rubber Balloon

On a cold winter day, you buy a rubber balloon on a street and then enter a warm room. Look at the balloon. During the heating the gas inside the balloon and its dimension increase until the temperatures have levelized. Certainly, the yarn must firmly close the opening of the rubber to maintain the mass of a gas inside strictly constant. The increase of balloon volume, the expansion of its constant mass of gas in thermodynamics is referred to as "process" between the initial (cold) and final (warm) state of gas in the balloon. Tension of the expanded film counterbalances internal pressure. The gas produces the work A=N.DF, where N= the force of tension per unit length, DF= the surface of rubber increase.

How do we calculate the value of A?

According to First Law, the change of internal energy ΔU equals heat input Q minus the work of gas expansion. This work isn't rubber expansion (elongation) only. It involves the shift of some air (volume ΔV) out of the room:

$$\Delta U=Q-p\Delta V=Q-(p-p_o)\Delta V-p_o\Delta V=Q-A-p_o\Delta V \tag{2.1}$$

If the gas in the balloon were heated somewhere in space without atmosphere, the expansion work would have been measured by rubber tension only. However, in our case, like in all the energy supply practice the environment exists around the considered object, it is very important restriction of ability to get work. The heat input into the balloon is $Q=T_o \cdot \Delta S_o$, where ΔS_o=the entropy decrease of air in the room. Aside from that there exists entropy gain in the balloon, as in any real process, hence ΔS exceeds ΔS_o. Formulating it we get:

$$-A=\Delta U+p_o\Delta V-T_o\Delta S=\Delta(U+p_oV-T_oS) \tag{2.2}$$

If the balloon assumed to be absolutely rigid, the film can't elongate, and no work might be done, A=0. Entropy can only rise, therefore the function in brackets decreases in any real process. In equilibrium it gets minimum.

EXERGY DEFINITION

The first mention of this function appeared in the paper of outstanding American physicist J. Gibbs in 1873, then it was repeated many times. It was given in the most rigorous form in a course of Theoretical Physics L. Landau and E. Lifshits (v.5, Statistical Physics, Sec. 20, Maximal work which might be done by a body in a given surrounding).

These are the words of Gibbs (in reversed translation), "If we want to express in one equation the necessary and sufficient condition of thermodynamic equilibrium of a substance, when it is surrounded by medium of constant pressure p_o and temperature T_o, this equation might be written down in the form

$$\Delta(V+p_oV-T_oS)=0 \tag{2.3}$$

where Δ means a variation due to change of arbitrary parts of a body.

The stable equilibrium condition requires the minimum of quantity in brackets."

The more protracted explanation we see in the Landau-Lifshits book: "Let a body be in the surrounding substance, which temperature T_o and pressure p_o differ from T and p of the body. The body can do the work on an object which might be thermally insulated from environment as well as from the body given. The substance with the body in it and with the work object forms a closed system.

The volume and energy of a substance are so large that no change in it occurs when arbitrary processes in the body take place. Hence, the temperature and pressure of the surrounding substance could be considered as constants...

The total change of body energy ΔU consists of three parts: the outer work A, the work done by substance and the obtained heat:

$$\Delta U=A+p_o\Delta V_o-T_o\Delta S_o \tag{2.4}$$

The total volume of substance and body is constant, hence $\Delta V_o=\Delta V$. According to entropy increase law $\Delta S+\Delta S_o \geq 0$ (entropy of the work source does not change) therefore $\Delta S_o>-\Delta S$. Hence

$$A \geq \Delta U-T_o\Delta S+p_o\Delta V \tag{2.5}$$

$$A_{max}=-\Delta(U+p_oV-T_oS) \tag{2.6}$$

Let us assume that the body in outer substance is alone and no work on it is done. Inside this body the irreversible processes will happen leading to equilibrium. In the mentioned inequality, the work A=0, therefore

$$\Delta(U+p_oV-T_oS)\leq 0 \qquad (2.7)$$

This means that the quantity $V+p_oV-T_oS$ will decrease due to body processes and in equilibrium it gets minimum."

The reader can see here the repetition of some words of the example with the rubber balloon. It is not coincidence, it is the two forms of the explanation: graphic and general.

The discovered function remained without a special name. Many papers used to call it "technical availability" or simply "availability". Since 1956, it has been referred to as "exergy," which is commonly used now.

At first it was specified for a closed system with only non-flow boundaries. Later it was generalized for a flow system and has become a universal value of usefulness of energy. For the last two decades, a large amount of books and papers has been devoted to exergy. Dr. Goran Wall in Sweden has collected nearly 2000 papers up until 1986. One being the handbook on exergy calculation recently published in Ukraine (Brodiansky 1991). See also Fratscher et al. (1988) Szargut et al. (1988), Tsatsaronis (1990).

Exergy concept is discussed in all the modern textbooks on Engineering Thermodynamics (e.g. Bejan, 1988). For professionals, I can recommend the very informative volumes of recent International Conferences ATHENS '91, ECOS '92 in Zaragoza, Spain and ENSEC '93 in Cracow, Poland (Editor-in-Chief George Tsatsaronis).

When reading the modern textbooks, we have to remember some well documented proposals in Russia.

On the 3rd All-Union Congress of Thermal Engineering, November 10-18, 1926 (Proc. v.III part I) G.A. Arons recommended "...the evaluation of rejected heat not by amount of calories, but by power which it may give passing through an ideal steam engine... Sooner or later the principle of thermodynamics $\eta=(T_1-T_2)/T_2$ where η is efficiency for boiler houses should be used, T_1 and T_2 are the temperatures of steam and of environment."

G.J. Fux in 1940, in the paper, New Method of Evaluation of Thermal Engine Performance, Energy Conference in Tomsk Industrial Institute, June 1940, "Red Flag" Publ. Tomsk 1941, proposed for the combined production of power and heat the formula:

$$\mu=\frac{E_{gain}}{E_{\max}}=\frac{860N+\sum Q(1-T_o/T_{av})}{BU_o} \qquad (2.8)$$

Here again, E_{max}=the maximum energy yield in the device, N=electrical power, Q=heat of certain temperature, T_o=environmental temperature, $T_{av}=Q/\Delta S$=averaged thermodynamic temperature, B=consumed fuel flowrate, U_o=maximal caloric value of combustion.

The above mentioned examples show that exergy is not a new idea.

Exergy calculations

According to modern views, the exergy of a thermodynamic system means the maximum work, which might give the system's process, leading to full equilibrium with the environment. The expression for gases is:

$$\varepsilon = H - T_0 S + RT_0 \Sigma X_i \ln X_i \tag{2.9}$$

where H=enthalpy, R=gas constant, X_i=molar parts of the given substances.

Keeping in mind the enthalpy definition, H=U+pV, the difference of exergy of a closed (non-flow) and open (flow) systems is $\Delta\varepsilon=(p-p_0)V$.

If macroscopic motion of a body as a whole is taken into account in the gravity field the exergy becomes

$$\varepsilon = H - H_o + \rho\left(\frac{V^2}{2} + gZ\right) - T_o(S - S_o) \tag{2.10}$$

Here g=gravity acceleration, Z=height.

The simplest form of exergy is for ideal gas:

$$\Delta\varepsilon = \int_{T_o}^{T} (1 - \frac{T_o}{T}) C_p dT + RT_o \ln P / P_o \tag{2.11}$$

in this case

$$\Delta H = \int_{T_o}^{T} C_p dT \approx C_p (T - T_o) \tag{2.12}$$

Total equilibrium with environment means the lack of any flow and equality of all the potentials.

The outer substance might be at a non-equilibrium state itself in real conditions around the considered object. As usual, the atmosphere, big water basins or earth core are assumed as reference substances in exergy calculations.

However, an engine might work in a restricted room, significantly changing the atmosphere, its chemical composition and temperature. The ocean waters might differ by temperatures so great that its difference is used to receiving power. The earth core contains some substances which are quite non-equilibrium with respect to atmosphere (fossil fuels).

At most, we can consider the atmosphere as the reference substance in exergy calculations. The choice is not easy in the case of chemical reactions.

Some substances obey seasonal variations and exergy depends upon it. Exergy of ice in winter is zero (in cold countries) but is quite high in summer. To get ice in summer a lot of work is needed.

The exergy of water in the cooling towers of industrial plants is zero in summer, but in winter it is considerable and it is used for heating the ventilation air in contact apparatuses (heat-exchangers).

The principal difficulty appears when we are trying to avoid atmosphere in combustion product release. Here we had better to select as reference condition either deep ocean layers or the Earth crust. It is the subject of contemporary investigations (see Szargut et al.).

In any case the outer substance which is considered as the reference state has to be specified.

The equilibrium might be partial, not full. The thin soft film boundary is assumed sometimes in between the system and atmosphere. In this case the temperatures and pressures are equal inside and outside the film, but chemical reactions are impossible as well as the mixing. The part of total exergy in the case of the above mentioned restricted equilibrium is referred to as "physical exergy." When chemical interaction with the reference substance exists, the exergy is called "chemical".

Chemical exergy

Exergy of a chemical composition MXy is to be calculated as the sum:

$$\varepsilon(MX_y)=\Delta G(MX_y)+\varepsilon(M)+y\varepsilon(X) \tag{2.13}$$

The first term in rhs is Gibbs energy of chemical reaction, second and third terms are exergies of components. As a rule, all of the date needed are available in handbooks.

The exergy of chemical elements is presented in the form of Mendeleev's periodical system (S. Yamauchi and K. Fueki, 1980) Fig. 2.1.

The simple example of chemical exergy calculations is methane exergy:

$$\varepsilon(CH_4)=\Delta G+\varepsilon(C)+2\varepsilon(H_2)=-51+410+2\cdot 235=829\text{kJ/mol}.$$

In the Fig. 2.1 aside from exergy and enthalpy, the reference chemical composition are presented. All the exergies of compounds might be calculated. One of the good handbooks is Stepanov (1985). (See also Losano and Valero, 1988.)

	Ia	IIa	IIa	IVa	Va	VIa	VIIa	VIII	VIII
1	4 **H** 28.11 34.16 (H_2O)								
2	27 **Li** 88.90 ($LiCl{\cdot}H_2O$) 97.58 ($LiCl$)	47 **Be** 142.03 ($BeO{\cdot}Al_2O_3$) 153.65 ($BeSO_4{\cdot}4H_2O$)							
3	9 **Na** 86.23 111.83 ($NaNO_3$)	11 **Mg** 147.26 173.49 ($CaCO_3{\cdot}MgCO_3$)							
4	10 **K** 92.46 118.22 (KNO_3)	8 **Ca** 170.26 194.41 ($CaCO_3$)	50 **Sc** 216.72 228.11 (Sc_2O_3)	12 **Ti** 211.66 225.80 (TiO_2)	23 **V** 168.47 185.3 (V_2O_5)	21 **Cr** 130.84 ($K_2Cr_2O_7$) 136.2 (Cr_2O_3)	14 **Mn** 110.24 124.29 (MnO_2)	7 **Fe** 87.99 (Fe_2O_3 170.37 ($Fe(NO_3)_3{\cdot}9H_2O$)	29 **Co** 68.93 ($CoFe_2O_4$) 95.71 ($CoCl_2{\cdot}6H_2O$)
5	18 **Rb** 93.11 118.32 ($RbNo_3$)	22 **Sr** 184.31 ($SrCl_2{\cdot}2H_2O$) 241.74 ($Sr(NO_3)_2$ $4H_2O$)	32 **Y** 222.86 ($Y(OH)_3$) 228.48 (YF_3)	20 **Zr** 253.01 268.21 ($ZrSiO_4$)	34 **Nb** 209.87 227.0 (Nb_2O_5)	37 **Mo** 170.75 173.99 ($CaMoO_4$)	**Tc**	73 **Ru** 0.0 0.0 (Ru)	79 **Rh** 0.0 0.0 (Rh)
6	42 **Cs** 93.43 105.84 ($CsCl$)	19 **Ba** 187.42 237.11 ($Ba(NO_3)_2$)	35 **La** 234.84 281.50 ($LaCl_3{\cdot}7H_2O$)	52 **Hf** 244.56 273.60 (HfO_2)	40 **Ta** 227.22 244.5 (Ta_2O_5)	26 **W** 195.56 198.79 ($CaWO_4$)	**Re**	77 **Os** 71.021 94.2 (OsO_4)	80 **Ir** 0.0 0.0 (Ir)

Legend (Cr example):

21 **Cr**	
130.84	Standard Exergy (kcal./mol.)
($K_2Cr_2O_7$)	Reference Substance for Exergy
136.2	Standard Enthalpy (kcal./mol.)
(Cr_2O_3)	Reference Substance for Enthalpy

	28 **Ce** 243.96 260.2 (CeO_2)	48 **Pr** 221.36 ($Pr(OH)_3$) 281.30 ($PrCl_3{\cdot}7H_2O$	33 **Nd** 231.13 277.11 ($NdCl_3{\cdot}6H_2O$)	**Pm**	44 **Sm** 230.13 276.11 ($SmCl_3{\cdot}6H_2O$)	58 **Eu** 208.53 255.11 ($EuCl_3{\cdot}6H_2O$)	45 **Gd** 229.03 275.11 ($GdCl_3{\cdot}6H_2O$)	59 **Tb** 226.43 273.51 ($TbCl_3{\cdot}6H_2O$)	51 **Dy** 229.03 276.11 ($DyCl_3{\cdot}6H_2O$)

VIII	Ib	IIb	IIb	IVb	Vb	VIb	VIIb	O
								72 **He** 7.20 0.0 (He)
			41 **B** 145.86 159.08 (H_3BO_3)	3 **C** 98.12 94.05 (CO_2)	2 **N** 0.08 0 (air)	1 **O** 0.47 0 (air)	17 **F** 73.62 60.77 (Ca_{10} $(PO_4)_6 \cdot F_2$)	76 **Ne** 6.47 0.0 (Ne)
			6 **Al** 188.39 (Al_2O_3) 271.39 ($Al(NO_3)_3$ $\cdot 6H_2O$)	5 **Si** 203.81 217.72 (SiO_2)	15 **P** 206. 97 200.84 (Ca_3 $(PO_4)_2$)	16 **S** 144.07 152.38 ($CaSO_4$ $2H_2O$)	13 **Cl** 5.61 (NaCl) 0 (Cl_2)	54 **Ar** 2.79 0.0 (Ar)
24 **Ni** 58.19 92.78 ($NiCl_2 \cdot$ $6H_2O$)	25 **Cu** 34.37 ($Cu_4(OH)_6$ Cl_2) 59.67 ($CuCl_2 \cdot 2H_2O$)	31 **Zn** 80.65 141.41 ($Zn(NO_3)_2$ $6H_2O$)	39 **Ga** 118.59 130.15 (Ga_2O)	43 **Ge** 117.86 131.7 (GeO_2)	49 **As** 92.3 110.53 (As_2O_3)	70 **Se** 39.99 (SeO_2) 64.27 ($H_2SeO_4 \cdot$ H_2O)	46 **Br** 8.21 5.82 ($PtBr_2$)	**Kr**
71 **Pd** 0.0 0.0 (Pd)	69 **Ag** 20.63 30.37 (AgCl)	32 **Cd** 72.70 ($CdCl_2 \cdot$ 3_2H_2O) 109.06 ($Cd(NO_3)_2$)	68 **In** 98.57 110.64 (In_2O_3)	30 **Sn** 123.26 138.8 (SnO_2)	61 **Sb** 97.92 116,15 (Sb_2O_5)	78 **Te** 63.66 77.10 (TeO_2)	64 **I** 6.12 1.61 (KIO_3)	**Xe**
74 **Pt** 0.0 0.0 (Pt)	75 **Au** 0.0 0.0 (Au)	65 **Hg** 31.48 53.60 ($HgCl_2$)	63 **Tl** 40.56 (Tl_2O_4) 75.3 ($TlCl_3$)	36 **Pb** 80.61 (PbCl (OH)) 85.9 ($PbCl_2$)	67 **Bi** 70.92 (BiOCl) 90.6 ($BiCl_3$)	**Po**	**At**	**Ru**

57 **Ho** 231.03 278.01 ($HoCl_3$ $\cdot 6H_2O$)	55 **Er** 229.63 277.11 ($ErCl_3$ $\cdot 6H_2O$)	66 **Tm** 213.74 225.70 (Tm_2O_3)	55 **Yb** 223.64 270.31 ($YbCl_3$ $\cdot 6H_2O$)	60 **Lu** 219.33 266.71 ($LuCl_3$ $\cdot 6H_2O$)		38 **Th** 278.41 293.12 (ThO_2)	53 **U** 268.21 284.8 (u_3O_2)	

Figure 2.1

However, a variety of reference substances leads to discrepancy of results for different investigators.

The possible discrepancy is evident from the Table 1, where exergies of chemical elements are commensurated. The date of Fueki and Yamauchi could be compared to that of Szargut and Stepanov. By and large they are in agreement within acceptable accuracy of engineering calculations. The significant difference for Si, Mn and Sn, where reference substances are the same, is less 7%.

Chemical exergy of fuels practically coincides to high caloric value. In the case of low grade fuels, the exergy exceeds by 10-15%. In the case of low-grade gaseous fuels (blast furnaces or coke oven), the difference disappears (See Baehr).

It has to be underlined that exergy does not always differ from energy. If no statistical quantities are involved in the considered process, it means no chaotic motions, in other words, if the number of possible states equals unity and its logarithm (the entropy) is zero, then exergy and energy are equal.

In mechanics of bodies, electromagnetism, wave transfer, the systems could be considered as quite deterministic. Here are no place for entropy and exergy does not differ from energy.

However in thermal and chemical processes, where innumerous particles move and some order might be seen through the law of chaos, the entropy plays the protagonist role and energy strongly differs from exergy.

Sometimes the exergy is referred to as an energy quality. The more exergy part of energy the better. It might be admissible if we restrict ourselves to the case of $T>T_O$ only. Otherwise, in the case when $T<T_O$ (widely used in refrigeration and cryogenics) exergy might exceed energy many-fold and can't be its part, measuring its quality.

Let us consider what happens if we deal with the cold fluxes, when the temperature of flowing heat Q is less than environmental. Here exergy flow is

$$\dot{\varepsilon} = \dot{Q} \cdot \frac{T - T_o}{T} \tag{2.14}$$

Exergy flow may be increasing when we try to get absolute zero. When temperature is elevated to that of its surrounding, the exergy flow disappears.

Exergy vector

Now it's the time to present the simple form of exergy vector, similar to (1.1), for steady-state process, without gravity and kinetic energy:

$$\delta_\varepsilon = \varphi j_q + G_\varepsilon \rho V + P \cdot V + (T - T_O) j_s \tag{2.15}$$

The difference from (1.1) in two points only: here G_ε (chemical exergy) replaced the G and in the last term the $(T-T_o)$ instead of T.

The simplest comparison of energy and exergy can be seen when considering the heat transfer process through the wall. Here the last terms of (1.1) and (2.15) are needed, see Fig. 2.2.

In the case of $T>T_o$ (power process or heating)the temperature is falling from left to right, in the direction of the energy current δ. The δ is the same on both sides of the wall according to First Law, or Oumov theorem: here must be $\mathrm{di}\boldsymbol{\delta}=\frac{\partial\delta_x}{\partial x}+\frac{\partial\delta_y}{\partial y}+\frac{\partial\delta_z}{\partial z}=0$. In this one-dimensional distribution nothing changes in y and z directions, hence $\frac{\partial}{\partial y}=\frac{\partial}{\partial z}=0$. It gives unavoidably $\frac{\partial\delta_x}{\partial x}=0$, therefore $\delta_x(x)$=const.

As $\delta_x=Tj_s$ and T is falling, j_s must increase from left to right. We used to say that the birth of entropy exists here. The exergy current has to decrease, due to the constant first and increased discarded term. The direction of energy and exergy currents is the same.

In the case of refrigeration, for instance, a freezing chamber wall, $T<T_o$ something differs (see right side of Fig. 2.2). The direction of energy and entropy flow (δ and j_s) is the same as before. However, exergy flow is inverted according to the sign of the term $(T-T_o)$. In the case of refrigeration, the energy and exergy currents are opposite.

In the last century there were two neighbors: the brewer and the butcher. The room of the butcher was located between the two rooms of the brewer. In one room a refrigerating machine has been installed and in another, the vessel containing a beer semiproduct, which must be cooled, was left. The pipes from the refrigerator with very cold brines were going through the butchers room. The butcher recognized that the pipes are always under a layer of snow and have begun to use it as a free cooler to store the fresh meat.

However, it spoiled the brewer's technology where the cold is needed and he complained to the municipal court that something was stolen. "What was stolen," asked the judge, "my energy," answered the brewer. But the butcher in reply told that he is the sponsor instead of a thief, because he gave his energy from the hot meat to the cold pipe, and won the process.

If the brewer knew what exergy was and was able to see the direction of the exergy current on the Fig. 2.2, he would have been the winner.

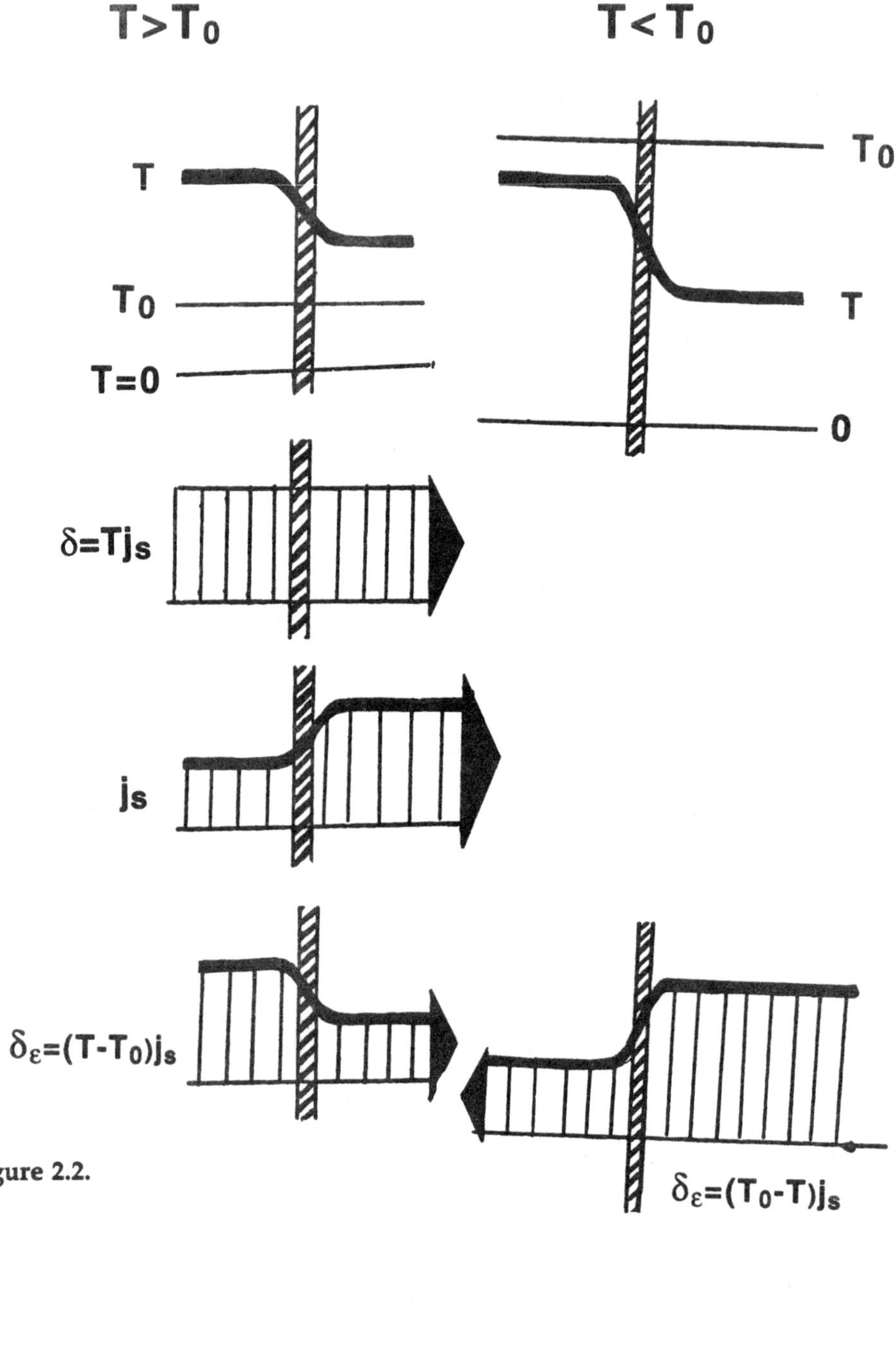

Figure 2.2.

Exergy losses

The stealing of exergy is a very rare occasion, whereas the losses of exergy are the natural event accompanying events of all energy transfer and conversion processes.

Where exergy loses?

At first, in the same places where energy is lost, where flows of substances, carrying energy exist, which are going out of the system. They are referred to as external exergy losses. Its value differs from the value of energy.

However the most important and less evident are the internal exergy losses due to generalized friction: mechanical friction, electrical one, mixing of flows and the less known but very important thermal friction, when the heat flows through large temperature difference. The last kind of friction is evident in Fig. 2.2.

If we only consider the heat flow $\dot{Q} = T\dot{S}$ we may ask what amount of mechanical power might be obtained from this heat. Keeping in mind that maximum work is measured by exergy of a system and maximum power by exergy flow we may compare last terms of δ_ε and δ, namely Tj_S and $(T-T_O)j_S$.

The relation of these terms

$$\eta = \frac{T - T_o}{T} = 1 - \frac{T_o}{T}$$

is a well known Carnot factor or "exergetic temperature function". It is the efficiency of ideal engine, where no entropy is gained and exergy losses are absent.

Let us consider some illustrative figures of exergy current densities.

Firebox of a boiler	10 MW/sq.m
Active steam turbine	100 MW/sq.m
Natural gas pipelines	50 GW/sq.m
High voltage line	1000 GW/sq.m
Modern laser	3,000 000 GW/sq.m

In the above mentioned examples, the exergy currents were nearly equal to the energy ones. The next two examples are related to the heat processes where they differ.

The first, "black smokers" on the ocean floor, where exergy current density seems to be the world champion of renewable energy sources, namely 1 GW/sq.m. (see p. 68)

The second, exergy current in a steam turbine condenser, where heat current 10kW/sq.m, T=315K and T_O=300K. Exergy current, 0,5kW/sq.m is 20 times less than the heat current density.

Here we see, that exergy current density may differ by many orders of magnitude. It appears as quite evident, that the high energy density in energy transfer devices, like a gas pipeline or electric line, would be impossible if it is accompanied by large exergy losses. On the contrary, the low current density use is forced by the designer to lessen these losses to an acceptable minimum. It especially relates to heat exchangers like to boilers.

The already known equations let us evaluate the intensity of exergy losses.

We have seen, that in steady-state processes according to First Law $div\delta=0$. Applying div to all the terms of (1.1), and bearing in mind that according to conservation of substance and charge div $(\rho V)=div j_q=0$, we can get entropy birth intensity:

$$-T\sigma_s=j_q \cdot grad\varphi+\rho\mathbf{V} \cdot gradG+div(P \cdot \mathbf{V})+j_s dradT. \tag{2.16}$$

Here the first term of rhs is electrical friction, the second one reflects "chemical friction" (mixing), the third is mechanical friction and the last, very important term presents "thermal friction," the entropy birth by entropy current, when heat is flowing in the body with nonuniform temperature (grad $T\neq0$).

In the last process $\sigma_s=-(j_s \cdot gradT)/T$. If the Fourier law for the process of thermal conductivity is valid, which we always assume, $Tj_s=-\lambda gradT$, where λ=const, then the entropy birth intensity is $\sigma_s=\lambda(gradT/T)^2$.

We see the sharper temperature change the greater entropy birth and exergy decreasing.

Conduction and Convection of Exergy

Here we continue the discussion of exergy currents. As before, this discussion is restricted within the cases of stationary flows. Practical thermal engineering often deals with stationary flows, which differ from the equilibrium state of classical thermodynamics that exist only as limiting cases.

The total energy of a physical system is the sum of the energies of the media and fields that compose the total system. Energy carriers are substances, charge, impulse and entropy, which are capable of transferring energy. Convection involves energy transfer by a moving substance. Conduction is energy transfer relative to a substance and may occur even in a stationary medium.

The specific energy ρe of a physical system and the energy flux density vector δ through the surface around a system are related by (See deGroot):

$$\rho e=\rho V^2/2+(E^2+B^2)/2+\rho U, \tag{2.17}$$

$$\delta=(\rho V^2/2+\rho U)\mathbf{V}+P\cdot\mathbf{V}+(\mathbf{E}\times\mathbf{B})+j_{Q,} \tag{2.18}$$

j_Q=heat flux density.

If the system includes *n* different chemical species, which may be reacting with each other, we must add to the rhs of Eq. (2.17) the term

$$\rho G = \sum_{1}^{n} \rho_i G_i, \tag{2.19}$$

i.e., the average chemical potential (Gibbs energy), similarly, on the rhs of Eq. 2.18 appears the term $\rho GV + \sum_1^n j_i G_i$ (j_i=diffusion flux density, ρ and V=averaged density and velocity of the mixture, respectively).

Recently, some authors (see Kreuzer) have divided the energy fluxes into conductive and convective parts, which is similar to the divisions made for electrical charge flows.

In electrodynamics, the current density is

$$\mathbf{j_q}=\sigma\mathbf{E}+q\mathbf{V} \tag{2.20}$$

where σ=conductivity and q=space charge. According to charge conservation,

$$\partial q/\partial t+\nabla\cdot \mathbf{j_q}=0. \tag{2.21}$$

The first component on the rhs of Eq. 2.20 represents the conduction current and the second is the convective current. These components are equal in their ability to create a magnetic field around the current. In a similar manner, we may note that energy flux density contains a term $\rho\mathbf{V}[(V^2/2)+U]$ for the convection of kinetic and internal energy, while the term for the conduction of mechanical energy is $P \cdot \mathbf{V}$. The vector $\mathbf{j_Q}$ is referred to as the conductive heat-flux density, whereas the convective heat flux is $\rho\mathbf{V}U$. Here, $\mathbf{j_Q}$ is assumed to be proportional to the temperature gradient, i.e., $\mathbf{j_Q}=-\lambda\,\nabla T$.

If we consider a compressible medium (gas), the spherically symmetric pressure p arising from mechanical stresses is excluded, i.e.,

$$P = -p\delta_{ik}+\tau_{ik}, \tag{2.22}$$

where δ_{ik}=Kronecker delta and τ_{ik}=tangential stress tensor. The enthalpy is $h=U+p/\rho$. The convective heat flux is $\rho\mathbf{V}h$.

If we introduce the Pointing vector after Slepian (see p. 22) the correction term $\mathbf{E}\times\mathbf{H}+\nabla\times(\varphi\mathbf{H})$, use $T\mathbf{j_s}=\rho\mathbf{V}h-\lambda\nabla T$ and also introduce the mass-flow density $\mathbf{j_m}=\rho\mathbf{V}$, then Eq. 2.18 becomes for stationary conditions without polarization

$$\delta=\mathbf{j_m}(V^2/2+G)+\varphi\mathbf{j_q}+\tau_{ik}\cdot\mathbf{V}+T\mathbf{j_s}. \tag{2.23}$$

The exergy-flow density vector now becomes

$$\delta_\varepsilon = j_m(V^2/2 + G_\varepsilon) + \varphi j_q + \tau_{ik} \cdot V + (T - T_o) j_s. \tag{2.24}$$

Diagrams for exergy transfer from one carrier to another, both with diminishing exergy flux and with constant energy flux, are presented in Chapter 3.

To continue the analogy between electrodynamics and non-equilibrium thermodynamics, we note the equation $\nabla \cdot \delta = 0$ for stationary energy flows, i.e., the energy vector lines are closed. Such a solenoidal vector may be represented by

$$\delta = \nabla\phi \times \nabla\Psi \tag{2.25}$$

where ϕ and Ψ are generalized potentials.

Equation 2.25 is similar to the equation for the Pointing vector. It describes a real energy flow by different carriers. If ϕ and Ψ are, respectively, the electrical and magnetic potentials, then $\delta = \mathbf{E} \times \mathbf{H}$.

For energy transfer involving non-compressible liquids or solids, a more convenient vector δ is used, which is of the form

$$\delta = j_m[V^2/2 + U + G - (\mathbf{E} \cdot \mathbf{P} + \mathbf{H} \cdot \mathbf{M})/2\rho] + \varphi j_q + P \cdot V + T j_s \tag{2.26}$$

we also use

$$\delta_\varepsilon = j_m[V2/2 + U + G_\varepsilon - (\mathbf{E} \cdot \mathbf{P} + \mathbf{H} \cdot \mathbf{M})/2\rho - T_o S] + \varphi j_q + P \cdot \mathbf{V} + (T - T_o) j_s. \tag{2.27}$$

In our proposed terminology, the first term represents energy convection by a moving medium and the last three terms conductive energy transfer as follows: electrical conduction φj_q, thermal conduction $T j_s$ and mechanical conduction $P \cdot V$.

The sum of the electromagnetic energy in the field and in the medium is given by

$$\varepsilon_o(E^2/2) + \mu_o(H^2/2) + (E \cdot P + H \cdot M)/2. \tag{2.28}$$

The last term is only transferred by moving media.

The thermal energy flux consists of two parts: a convective part $j_m U$ and a conductive part $-\lambda \nabla T$.

When we refer to thermal-energy production, convective energy flow is in joules or calories and is transferred by steam or hot water flow. There are no physical contradictions and the term thermal-energy production is fully analogous to electrical energy production.

The role of the velocity $\mathbf{V}$ in the term $P \cdot \mathbf{V}$ of Eq. (2.26) is unusual. It is an intensive property (potential) whereas P (the impulse flow) is an extensive property. Thus, $\delta = P \cdot \mathbf{V}$ describes mechanical work flow, for example, through a shaft from a turbine to a generator or by pulling a belt.

The concept of power production as energy transfer from one carrier to another (at a power station, where we pass from fuel flow to entropy flow in the boiler, then to impulse flow in a turbine and, finally to charge flow in the electrical line) is useful for optimization of exergy carriers' flow density.

The next table presents the conductive and convective part of exergy flows.

Summary of symbols and thermodynamic quantities: φ=electrical potential, G_ε=chemical exergy, P=polarization, M=magnetization, λ=thermal conductivity.

Quantity	Stationary Flows	
	Conductive Part	Convective Part
Electrical charge j_q	$\sigma \mathbf{E}$	$q\mathbf{V}$
Thermal energy Tj_s	$-\lambda \nabla T$	$\rho\mathbf{V}(U+p/\rho)$
Mechanical impulse	$\tau_{ik}=p\delta_{ik}$	$\rho V_i V_k$
Mechanical energy	$(\tau_{ik}-p\delta_{ik})\cdot\mathbf{V}$	$\rho\mathbf{V}\cdot V^2/2$
Electromagnetic energy in media	$\varphi\sigma E$	$\mathbf{V}(\mathbf{E}\cdot\mathbf{P}+\mathbf{H}\cdot\mathbf{M})/2$
Chemical energy	$\Sigma G_i j_i$	$\rho\mathbf{V}G$
Total energy without diffusion	$\varphi j_q+(\tau_{ik}-p\delta_{ik})\cdot\mathbf{V}-\lambda\nabla T$	$\rho\mathbf{V}[V^2/2+U+G-$ $-(\mathbf{E}\cdot\mathbf{P}+\mathbf{H}\cdot\mathbf{M})/2\rho]$
Exergy δ_ε	$\varphi j_q+(\tau_{ik}-p\delta_{ik})\mathbf{V}-$ $-\lambda\nabla T(T-T_o)/T$	$\rho\mathbf{V}[V^2/2+U+G_\varepsilon-T_oS-$ $-(\mathbf{E}\cdot\mathbf{P}+\mathbf{H}\cdot\mathbf{M})/2\rho]$

Very often on scientific gatherings, the questions what is work and what is heat arise.

The writer frankly speaks that he doesn't know what energy is. But to the people who think that they know what it is, we can offer such simple definitions:

Work= energy, conductively carried by impulse

heat= energy, conductively carried by entropy.

$$\text{Work current density}=P_{ik}\cdot V=\delta^W \tag{2.28}$$

Heat current density=$-\lambda\nabla T=\delta^Q$ (2.29)

Work flux= $\int_F \delta^w \cdot dF = \dot{W}$ (2.30)

Heat flux= $\int_F \delta^Q \cdot dF = \dot{Q}$ (2.31)

Total amounts of Work W= $\int \dot{W} dt$ (2.32)

Heat Q= $\int \dot{Q} dt$ (2.33)

The convective part of energy currents reflects solely energy transfer, because transferring quantity is the function of a point, it does not depend on the path of processes.

General forms

For the most general case of nonsteady flows, if the exergy per unit of volume is

$$\rho\varepsilon = \frac{1}{2}\rho V^2 + \frac{1}{2}(\varepsilon_o E^2 + \mu_o H^2) + \rho U + \rho G_\varepsilon - T_o \Delta S \tag{2.34}$$

the corresponding exergy flux density vector is

$$\mathbf{\delta}_\varepsilon = (\frac{1}{2}V^2 + U + G_\varepsilon)\rho\mathbf{V} + \mathbf{E}\times\mathbf{H} + P\cdot\mathbf{V} + \mathbf{j}_Q - T_o\mathbf{j}_s \tag{2.35}$$

For electromagnetic field energy we have well known Pointing theorem

$$\frac{\partial}{\partial t}(\varepsilon_o E^2 + \mu_o H^2)\cdot\frac{1}{2} + div(\mathbf{E}\times\mathbf{H}) = -j^2/\sigma \tag{2.36}$$

here, rhs is Joule heating, leading the field energy to disappear. For specific exergy we know

$$\frac{\partial}{\partial t}(\rho\varepsilon) + div\mathbf{\delta}_\varepsilon = -T_o\sigma_s \tag{2.37}$$

here rhs is always a negative intensity of lost work.

The total energy is conserving, the corresponding equation (1.2) has no rhs. Now we can stress on the analogy between the field energy and exergy, their governing equations (2.36) and (2.37) both have the negative rhs, leading to destroying.

For simplification some further algebra is needed. Substituting

$$\rho VU + j_Q = Tj_s \tag{2.38}$$

and modified expression for the Pointing vector after Slepian (see p. 29):

$$\mathbf{E}\times\mathbf{H} + rot(\varphi\mathbf{H}) = \varphi(j + \frac{\partial\mathbf{D}}{\partial t}) - \frac{\partial\mathbf{A}}{\partial t}\times\mathbf{H} \tag{2.39}$$

where $\frac{\partial\mathbf{D}}{\partial t}$=displacement current,

$$\delta_\varepsilon = (G_\varepsilon + \frac{1}{2}V^2)\rho V + P\cdot V + \varphi(j + \frac{\partial\mathbf{D}}{\partial t}) - \frac{\partial\mathbf{A}}{\partial t}\times H + (T - T_o)j_s \tag{2.40}$$

The above is the most general form of exergy vector. Some problems may appear when we deal with electromagnetic radiation, with large amount of different frequencies, like, for example, the light from solar radiation. The term $\frac{\partial A}{\partial t}\times H$ describes this process, however the radiation in this case contains some entropy. The exergy of radiation presents the special branch of exergy analysis. It is beyond the scope of this book.

As an appendix to this paragraph, we present this system of equations of the thermodynamics of continuum. We don't see any reason to discuss it, however, it might be useful in some controversial cases to refer back to:

Reference system of equations (without gravity and polarization):

$$D = \varepsilon_o E \quad \nabla\cdot D = q \quad j_q + \frac{\partial D}{\partial t} = \nabla\times H$$ *field equations*

$$\mathbf{B} = \mu_o H \quad \nabla\cdot B = 0 \quad \nabla\times E = -\frac{\partial\mathbf{B}}{\partial t}$$

$$\frac{\partial\rho}{\partial t} + \nabla\cdot(\rho V) = 0$$ *mass conservation*

$$\frac{\partial(\rho V)}{\partial t} + \nabla\cdot(\overbrace{p\delta_{ik} + \tau_{ik} + \rho V_i V_k}^{P_{ik}}) = qE + j_q\times B$$ *mechanical impulse*

$$\frac{\partial(D\times B)}{\partial t} + \nabla\cdot[\overbrace{E_i D_k + H_i B_k - \delta_{ik}(D\cdot E + B\cdot H)}^{M_{ik}}] =$$

$$= -(qE + j_q x B)$$ *electromagnetic impulse*

$$\frac{\partial(\rho V + D \times B)}{\partial t} + \nabla \cdot (P_{ik} + M_{ik}) = 0$$ *impulse conservation,*

$$\frac{\partial q}{\partial t} + \nabla \cdot j_q = 0$$ *charge conservation,*

$$\frac{\partial s}{\partial t} + \nabla \cdot j_s = \sigma_s \geq 0$$ *entropy rise,*

$$\frac{\partial}{\partial t}(U + \frac{P}{\rho} + G + \frac{V^2}{2} + \frac{H \cdot B}{2} + \frac{E \cdot D}{2}) +$$

$$+\nabla \cdot [\rho V(U + P/\rho + G + \frac{V^2}{2}) - \lambda \nabla T + V \cdot \tau_{ik} +$$

$$+\varphi(j_q + \frac{\partial D}{\partial t}) - \frac{\partial A}{\partial t} \times H = 0$$ *energy conservation*

Note here, that only the total impulse, the sum of mechanical and electromagnetic ones is conserving (the forces in rhs are equal and opposite).

Exergy and Informational Entropy

The objective of this paragraph is to connect all the information on the kinds of entropy, including informational entropy, with exergy. The papers by Kolmogorov, Ebeling and Zurek form the basis for this development which is an important part of exergonomics. W.H. Zurek has considered "a computerized engine consisting of a physical system in contact with a thermal reservoir at temperature T and controlled by a universal computer C. The system is initially described by some density matrix. Its standard ensemble entropy, measured in bits, is given by

$$H(\rho) = -\mathrm{Tr}\rho\log_2\rho. \tag{2.41}$$

The entropy of each outcome ($H\rho_k$) is smaller than $H(\rho)$ and the average decrease of entropy is equal to

$$[\Delta H] = H(\rho) - \sum_k P_k H(\rho_k) = -\sum_k P_k \log_2 P_k \tag{2.42}$$

The computer can convert this gain of information into useful work at no thermodynamic expense. Subsequently, useful work

$$\Delta W = T(H(\rho) - H)\rho_k)) = T\Delta H_k \tag{2.43}$$

could be extracted by allowing the constraint to relax in a slow isothermal fashion, so that ρ_k is transformed into ρ (Boltzmann constant $K_B=1$).

The net amount of extractable work is the sum of the usual ensemble entropy, which specifies the missing information, plus the algorithmic randomness of available information.

$$S=H+K \tag{2.44}$$

The process of energy extraction is discussed in terms of physical entropy defined as the sum of known disorder (measured by the algorithmic randomness) and remaining ignorance (measured by the Shannon entropy).

The increase in potential to do useful work, as a result of information gain, must be measured not just by the decrease in the ensemble entropy ΔH_k, but by the sum of ΔH_k and ΔK_k which gives the least numbers of bits needed to write down the outcome of the measurement.

The net amount of work that can be gained following measurement, with outcome K is

$$W_k=T(H_k+K_k)." \tag{2.45}$$

In another recent paper (Ebeling, Volkenstein) we see: "Inside a body *entropy has no form,* it is just plain entropy. However, passing from one body to another, it assumes a form. One of the possible forms is Shannon entropy, i.e., entropy processing information character. Due to this relation, information transfer is bound by the second law of thermodynamics. Information processing is possible only in systems which are pumped with high valued energy. What electrical energy means to the present day computers, food means to animals and radiation to plants. Information processing is connected with transfer of a special form of entropy called information entropy from one system to another or with the change of other forms of entropy into information entropy. The second law refers to the sum of all entropy forms as well as the first law refers to the sum of all energy forms only."

Our final quoting lists all the kinds of entropies (Ebeling et al.):

I) Phenomenological (thermodynamic) entropy,

II) Shannon entropy $H=-\int dx P(x) \ln P(x)$, related to the statistical entropy proposed by Boltzmann: $S=K_B H$(microstate).

III) Algorithmic entropy. Kolmogorov and Chaitin introduced in 1965 the concept of complexity of a sequence expressing the minimal length of programs which are able to generate this sequence.

IV) Mean entropy per symbol for sequence of letters: $H=\lim (H_{r+1}-H_r)$ where H_r denotes the Shannon entropies for r-words. The minimal value of H is zero, corresponding to regularity and complete predictability of the following letter in a sequence.

V) Kolmogorov entropy for dynamical systems, defined by $K=\mathrm{Sup}H(\{A\})$. The minimal value of K is zero which means regularity or periodicity of the motion which allows predictability of the future.

A positive value of K means chaotic motion and restricted predictability. The list of entropy definitions given above is not complete. Other measures of order are connected with the breaking of symmetry or with the increase of correlations."

The author's aim is to combine the above mentioned papers in order to generalize exergy:

$$\varepsilon = h - T_o S \tag{2.46}$$

where h=specific enthalpy, T_o=reference temperature.

Hitherto, only one form of entropy was used in the exergy formulation, a phenomenological one.

It is interesting to note the formula of Ebeling for the dimensionless actual value of the energy when entropy is decreased (here X=extensive parameter, E=fixed energy, t=time):

$$W(E,X,t) = 1 - \exp\{[S(E,X,t) - S_{eq}(E,X,t)]/K_B\} \tag{2.47}$$

If the difference between the two quantities of entropy (S) is small, this value function is proportional to $S-S_o$ as is sometimes common to ordinary exergy.

After considering all kinds of entropies and Zurek's pioneering formula, the formula for the generalized exergy might appear as:

$$\varepsilon = h - T_o \Sigma s_i \tag{2.48}$$

where Σs_i=the sum of all kinds of listed entropies.

We might question whether such a calculation is possible in principle, because of the very low value of informational entropy compared with the thermodynamical one. Here we have a very low amount of degrees of freedom, in contrast with innumerous positions of a particle.

As an example, let us consider the electrical heating of a flow in a small pipe. A medium is flowing through the pipe with a strict constant velocity V. Along the axis of the pipe, a straight wire is mounted, heated by the current I from a power source of voltage U(t). The enthalpy of the flowing medium is $h=C(T-T_o)$. The energy balance given for the convective energy current is;

$$\rho V F h = U^2/R = C \rho V F (T - T_o) \tag{2.49}$$

Here R=resistance, C=specific heat capacity, F=cross-section of the pipe. This equation shows the increase of enthalpy flow due to the power income of the wire.

The ability to do work and the exergy flow at the entrance of the device is equal to the electric power. However at the exit, the exergy gain is as follows:

$$\varepsilon = h - T_o \Delta s = C(T - T_o) - C(T - T_o) T_o / T = C(T - T_o)^2 / T. \tag{2.50}$$

Here s=Clausius entropy in heating:

$$s=h/T=C(T-T_o)/T \tag{2.51}$$

Hence keeping in mind (2.49)

$$\varepsilon=CU^4/T(RC\rho VF)^2=U^4/A \tag{2.52}$$

Here the deviation of T from T_o was assumed to be small; therefore, the denominator in (2.52) A=Const.

The voltage U(t) is changing; in order to transfer this information, the information entropy current has to be added to the above mentioned entropy, at least in the form of Kolmogorov-Chaitin entropy. By means of electrical heating it is possible to fix the information on a special thermally sensitive paper (telefax).

If the simple function like sin (kt) is to be transferred, the program is quite short. In the case of an arbitrary function it might be expanded in Fourier series:

$$F(t)=\sum_{k=0}^{N} A_k \cos kt + B_k \sin kt \tag{2.53}$$

The algorithmic complexity of the sequence of the terms $F(t_o)$, $F(t_1)\ldots F(t_n)$ does not depend upon the amount of harmonics in F(t). Instead, it depends upon the algorithmic complexity of the series of coefficients, A_kB_k, and the sequence of control switching moments $t_1, t_2,\ldots t_s$.

If the memory M of a computer is sufficient, the information entropy does not increase because information is accepted. If memory becomes overcrowded, then the entropy increases. This might be expressed in the form:

$$\Delta K=C(n-M) \tag{2.54}$$

where n=full amount of control switching, C=dimensional coefficient.

Therefore, the generalized exergy is as follows:

$$\Delta\varepsilon=\Delta h-T_o[\Delta S+C(n-M)] \tag{2.55}$$

The practical calculations by means of (2.55) are still doubtful. The problem seems to be formidable, the constant C still unknown. It is worth the efforts of new generations of researchers using computers and algorithmic theory.

Exergy-Enthalpy Diagrams

In any thermodynamic textbook, the state diagrams like temperature-entropy or enthalpy-entropy are widely used. Less known are exergy-enthalpy diagrams. We wish to present here such diagrams for the two most important substances, namely steam and carbon dioxide.

The steam diagram (Fig. 2.3) covers very wide pressure interval from 0,000618 MPa (0,00618 bar) to 100 MPa (1000 bar) in all the practical temperature interval O°C-800°C.

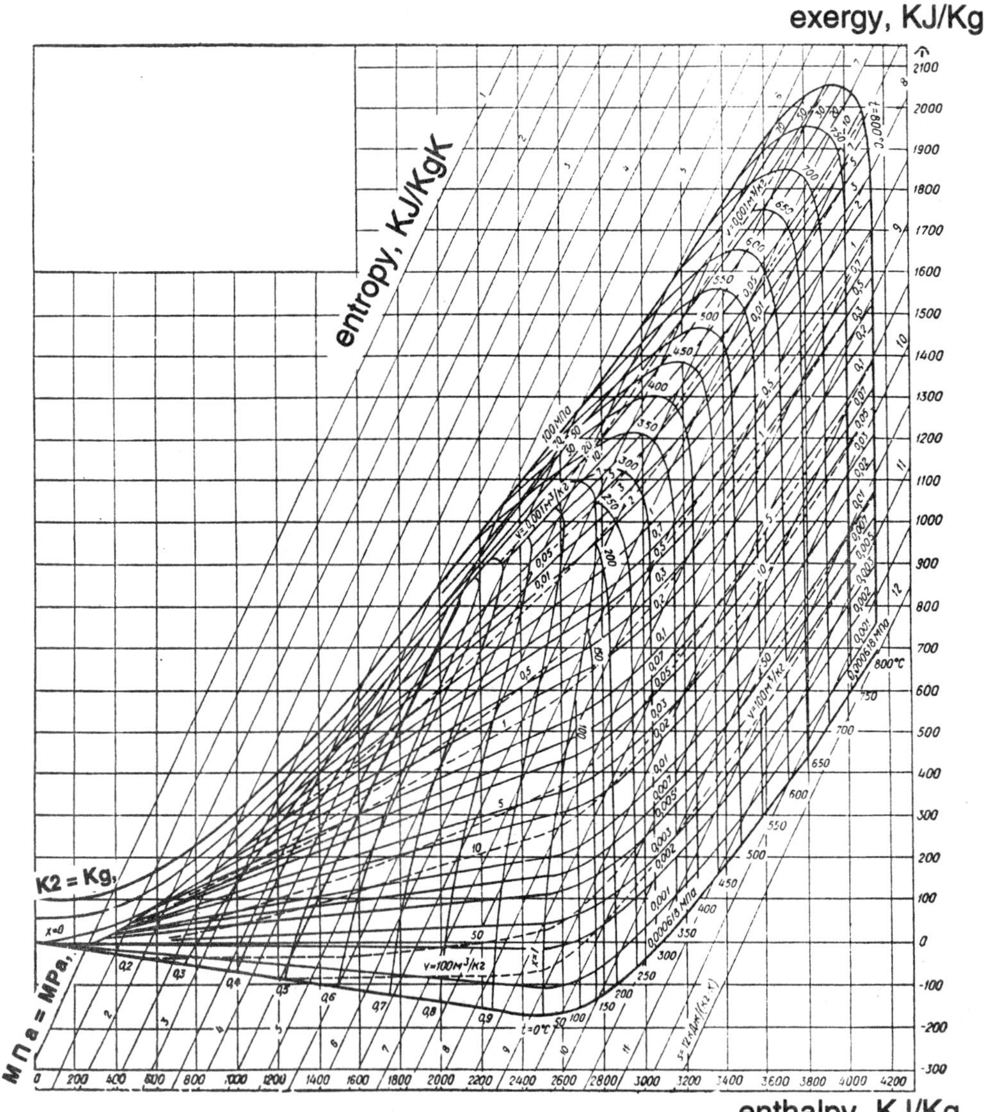

Figure 2.3 Exergy-enthalpy diagram for steam. T_0=293K

According to T_0S term in the exergy definition, the straight lines for entropy are presented from S=1kJ/kgK to 12kJ/kgK. Note a region of a negative values of exergy in this diagram.

The next diagram for carbon dioxide (Fig. 2.4) relates to a rather cold substance near the saturation line. The reference temperature here is nearly the same as for steam T_0=290K and the same position of entropy lines. The pressure interval covers 0,1-12 MPa (120 bars) and temperatures from –80°C to 130°C.

This diagram is of interest when we deal with the process of carbon dioxide liquefaction in the new clean power cycles of internal combustion without exhaust gases (see Chapter 6).

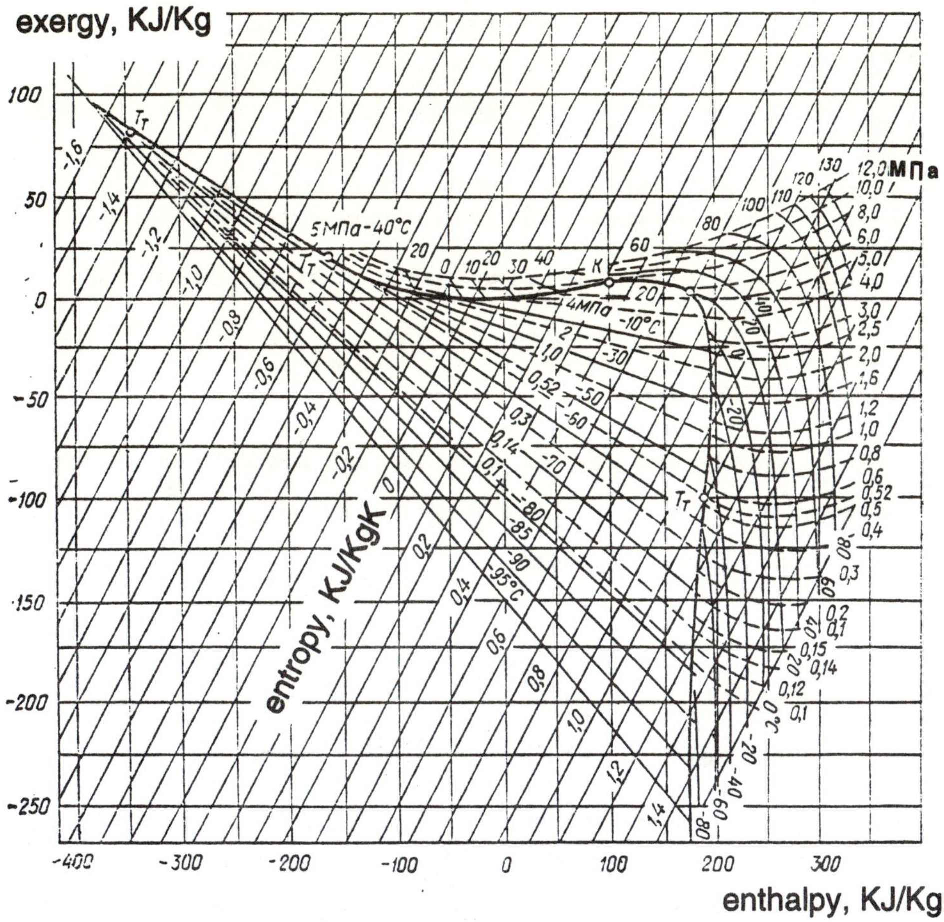

Figure 2.4 Exergy-enthalpy diagram for carbon dioxide near the saturation line.

Electric Power Current Density in a Water Heater

The greatest role of electricity in energy supply needs no explanation. It is the foundation of modern human life. Therefore, it might appear as a trick or even a sad joke, that the density of the electrical power current does have many expressions, many formulas and none of us know what is correct.

R. Feinman in his lectures gave the most common expressions for the electromagnetic energy in unit of volume W and energy vector δ:

$$W = \frac{\varepsilon}{2}E^2 + \frac{\mu}{2}H^2 \tag{2.56}$$

$$\delta = \mathbf{E} \times \mathbf{H} \tag{2.57}$$

and Pointing formula:

$$\frac{\partial W}{\partial t} + div\delta = -E \cdot j \tag{2.58}$$

saying that, in effect for W and δ there exists innumerous amount of different expressions and no one investigator tried experimentally to approve what is the truth.

To explain the setting, we have to remember that the electric field is

$$E = -\nabla\varphi - \frac{\partial A}{\partial t}; \qquad \nabla \times A = B \tag{2.59}$$

(it is an immediate solution of the Maxwell equation $\nabla \times E = \frac{\partial B}{\partial t}$). If we put it into δ, we see:

$$\begin{aligned} \delta = E \times H = -\nabla\varphi \times H - \frac{\partial A}{\partial t} \times H = \varphi\nabla \times H - \nabla \times (\varphi H) - \frac{\partial A}{\partial t} \times H = \\ = \varphi(j + \frac{\partial D}{\partial t}) - \frac{\partial A}{\partial t} \times H - \nabla \times (\varphi H). \end{aligned} \tag{2.60}$$

If we redefine energy vector as $\mathbf{E}\times\mathbf{H}+\nabla\times(\varphi\mathbf{H})=\delta'$ it gives:

$$\delta' = \varphi(j + \frac{\partial D}{\partial t}) - \frac{\partial A}{\partial t} \times H; \quad \delta' - \delta = \nabla \times (\varphi H) = rot(\varphi H) \tag{2.61}$$

Now we try to demonstrate the usefulness of δ′ vector (the Pointing vector with Slepian addition $\nabla\times(\varphi H)$) by means of simple example of electrical water heater.

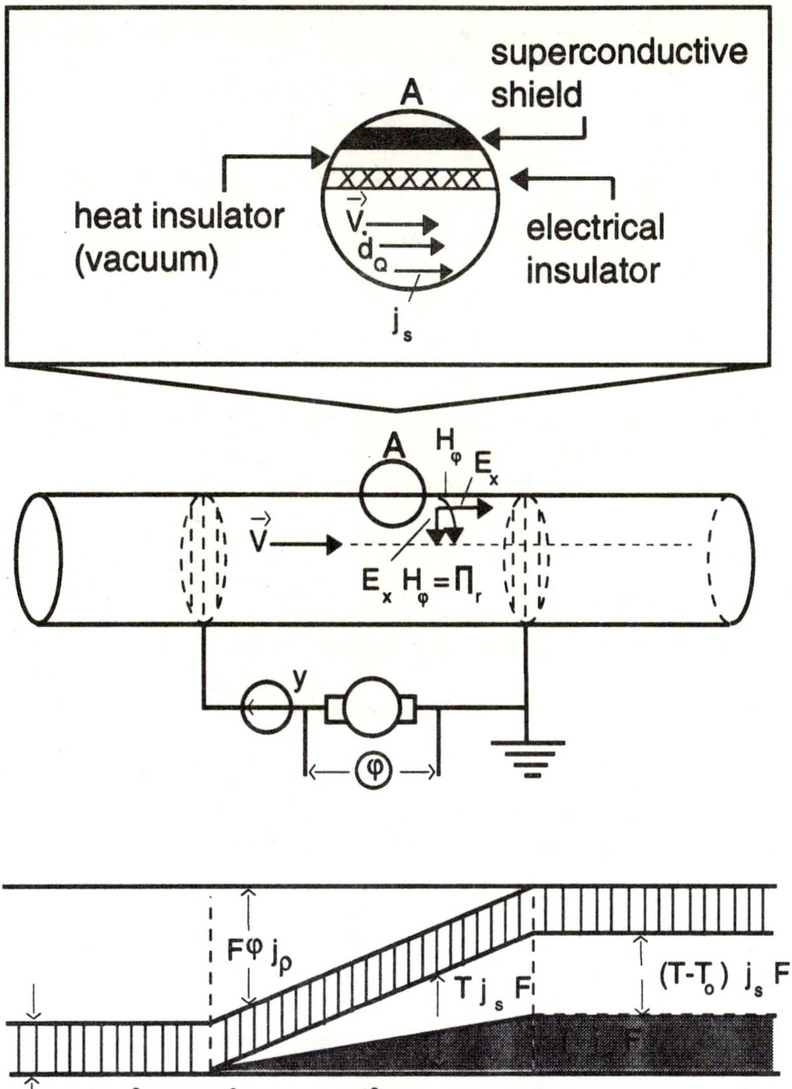

$$F j_M \frac{V^2}{2} = F\rho \frac{V^3}{2}; \quad \delta = j_M \frac{V^2}{2} + \varphi j_\rho + T j_s;$$

$$\delta_\varepsilon = j_M \frac{V^2}{2} + \varphi j_\rho + (T-T_o)\, j_s$$

Figure 2.5 Scheme and exergy flow diagram of an electric water heater

On the Fig. 2.5 we see such a heater, the pipe with flowing water, where the two grid electrode system is installed. The voltage φ between the grids maintains the DC electrical generator.

Let us assume (for educational aim only!) that around the pipe wall the triple ideal layers exist, namely superconductive, vacuum (ideal heat insulation) and resin (ideal electrical insulation).

The electric current j, flowing along the velocity of water V, causes Joule heating of water and its temperature increases.

What can we say on the flow of energy into this water? The electric field here has only longitudinal projection Ex, magnetic field only tangential Hφ its vector product $\delta = E \times H = E_x H_\varphi = \delta_r$, its direction is radial, to the axis of the pipe. Here we have to understand how this energy flow vector can overthrow the triple shielding around the pipe. It is by definition the ideal protection to any field.

The present writer has to speak frankly that in his modest practice of lecturing, the obstacle was insurmountable.

Quite another thing appears when we use the δ′ vector. In our strictly steady-state process $\left(\frac{\partial A}{\partial t} = \frac{\partial D}{\partial t} = 0\right)$ the energy vector is simply $\delta' = \varphi j$.

Energy is flowing along the electrical charge current. Here we can consider the electrical charge as a real energy (and exergy!) carrier.

Before the left electrode, the only exergy in water is a kinetic one, $\rho V \cdot \frac{V^2}{2}$. Along the pipe, it is constant unless water is boiling. The boiling is beyond our scope. After the left electrode, the exergy flow consists of three parts: mentioned kinetic one, the electric φj_q and the thermal one $(T-T_o)j_s$. After the right grid electrode, the electric part disappears and we have two constant exergy current components: kinetic and thermal. Note here the exergy loss current $T_o j_s$.

Let us compare the two final integral quantities; the total power consumption N to heat the water. In the case of δ use we need to multiply the vector $E_x \cdot H_\varphi$ over the pipe surface 2πRL. Keeping in mind that $E_x = \varphi / L$ and $H_\varphi = J/2\pi R$ we have:

$$N = \frac{\varphi}{L} \cdot \frac{J}{2\pi R} \cdot 2\pi RL = \varphi J \tag{2.62}$$

In the case of δ′ use, keeping in mind that $j = J/F$ where $F = \pi R^2$=pipe cross-section area, we get:

$$N = \varphi \cdot j \cdot F = \varphi J \tag{2.63}$$

The result is the same as in our previous case. We can safely say that the results always will be the same, because the difference between δ and δ' disappears in integration over the surface.

Here we see that the difference between two expressions δ and δ' is rotor of the product of electric potential φ and magnetic field H. According to the well known identity $\nabla \cdot \nabla \times(\quad)$=divrot $(\quad)$=0, in the Pointing theorem the δ and δ' are indiscernible. We have to keep in mind that we really only know the divergence in the Pointing theorem and nothing more about the energy vector.

Therefore, the expression of δ' might be used in practice along with the δ and it gives us some benefits, avoiding the misleading conclusion when we use δ. This was discovered by Slepian in 1942.

It is easily seen in this triple-shielding pipe that energy flow is constant along the pipe: no energy flow can enter through the triple wall. Therefore energy efficiency of this device is always $\eta_1=\delta_{out}/\delta_{in}=100\%$. Here δ=total energy vector. In contrast to this, exergy efficiency is less unity due to entropy rise inside the heater. In the domestic water heater, giving 60°C water this efficiency is nearly $\Delta T/T=60/60+293=20\%$ only.

We see in this simple example that the use of δ in exergy current and diagram treating needs the "jump" of this current over the triple shielding, which is for many of us the violence over the normal logic.

We know only one attempt to explain the difficulty in the ordinary Pointing vector use (Parton et al.). Let us consider the fixed charge capacitor in the permanent external magnet gap. The lines of electric and magnetic fields located there are crossed rectangularly. There exists vector **E**×**H**, but no energy currents.

The authors pointed out that in using **E**×**H** expression, we must ensure that **E** and **H** are casually related, that is, both are simultaneously from the same source. In the mentioned example, the electric field is from a charged capacitor and the magnetic field from a permanent magnet, that is from different sources. Hence, here must be **E**×**H**=0.

Continuing this logic, we have to remember that the magnetic field itself is the trace of moving electrical charges which form the electric current. The above mentioned explanation, in the case of electrical heater, is valid because magnetic field H_φ originated due to charges, moving between the electrodes. The electrical field is created by charges on the electrodes, just like in the capacitor, and here the **E**×**H**=δ as the energy current is invalid. We saw that the δ' in this example gives acceptable results from the point of general electromagnetic theory and convenient to practical explanation and educational aims as well.

Electrical Heating of Oil Wells

Just considered electrical water heater is merely an educational model. There exists, however, a particular case, where electrical heater might be extremely useful. It is the thermal enhancement of oil recovery, briefly described by Silvester.

Let us compare some methods of energy transfer in a tube of oil well. The diameter of this tube is around 150 mm, not more. Regardless to the kind of energy carrier we need to increase temperature in the well bottom and decrease viscosity of oil, it is the principal aim of energy transfer. For that matter we need to find energy carrier of highest energy current density without extra losses. Widely used is convection of energy by steam injection. By order of magnitude it might give for critical state of steam:

$$\delta=\rho Vh=315.10.2107=6.6.10^9 \text{ W/sc.m}$$

Here p=221 bar, t=374°C, h=2107 kJ/kg, ρ=315 kg/cub.m are used. The normative figures of steam generators are ρV=1500 kg/sq.m.s h=2800 kJ/kg, hence $\rho Vh=4.2.10^9$ W/sq.m, it coincides with former figures, based on critical steam parameters.

In case of superheated high pressure water $\delta=\rho Vh=732.10.1990=9.4.10^9$ W/sq.m.

If we transfer energy by liquid oxygen injection $\delta=\rho VH=1270.10.10^7$ $=1.27.10^{11}$ W/sq.m. Here the same velocity 10 m/s and H (enthalpy of combustion) 10 MJ/kg of oxygen are used.

We can try to transfer energy by means of hydraulic pressure, assuming pressure drop Δp=250 bar and the same velocity:

$$\delta=\Delta pV=250.10^5.10=2{,}5.10^8 \text{ W/sq.m.}$$

It is much less than above mentioned in case of steam, water or oxygen.

Finally we can use well insulated high voltage cable (φ=100 kV) with electric current density $\varphi=10^7$ A/sq.m. Here we have

$$\delta=\varphi j=10^5.10^7=10^{12} \text{ W/sq.m.}$$

It is the highest figure, unattainable by other energy carriers. In a long tube, say 1000 m or more, the losses of heat in case of steam or hot water are certainly very high as well as hydraulic friction in a tube, in many practical cases they are prohibitive. The use of oxygen for underground combustion has been forbidden in many countries as rather dangerous.

In case of electrical heating there are very small resistance losses (around 3%) and negligible leakage of current through the cable insulation.

Presented very simple comparison of energy current density gives us an idea of electrical heating for enhancement of oil recovery.

Chapter 3.

EXERGY EFFICIENCY

A clever number reflects all the kinds of sense.
N. Gumilev

General Formulations

In the simplest example, Fig. 2.2, we saw that on both sides of the wall in heat transfer the energy current is the same. The relation from right to left is unified and irrelevant to temperature change and entropy rise. This relation is the heat transfer efficiency based on First Law only. In other words we saw that energy balance efficiency is totally blind with respect to entropy rise. It is the inherent illness of the First Law efficiency.

However, in the heat transfer, the flow of heat might be cooled down to ambient (reference) temperature when energy is unable to do work. The same relates to the steam flow in a power plant condenser before condensation. Its enthalpy is quite large, but its ability to give power is nearly nil.

The exergy efficiency only presents the quality of considered energy flow. It is equal to the ratio of exit exergy flow to entrance one:

$$\eta_\varepsilon = \dot{\varepsilon}_{out}/\dot{\varepsilon}_{in} = 1 - \dot{D}/\dot{\varepsilon}_{in} \tag{3.1}$$

here $\dot{\varepsilon} = \int_F \dot{\delta}_\varepsilon \cdot dF$=exergy flux, if it is uniform $\varepsilon = \delta_\varepsilon F$, D=losses. Let us divide exergy flow on the three parts, namely exergy of substance flow ε^m, of heat ε^Q and of work (it is just work itself) ε^W. Then the exergy flow balance of any object:

$$(\varepsilon^m + \varepsilon^Q + \varepsilon^W)_{in} = (\varepsilon^m + \varepsilon^Q + \varepsilon^W)_{out} + D \tag{3.2}$$

Here $D = T_0 \Sigma \dot{S}_i$, the sum of entropy gains multiplied by reference temperature.

The external exergy losses are any parts of rhs flows deflected from consumer and released into the environment. The widely used term "secondary energy sources" means simply the released exergy.

In the hot discussions on future energy problems the secondary energy plays a role as a part of energy budget. But this role should not be overestimated. The secondary energy cannot be the base of energy supply of a country, just the same as the income for returned empty bottles cannot form a personal budget. The Second Law prevents energy recycling (in contrast to substance recycling) and will force mankind to use external exergy flows.

The reasons of the decline in exergy efficiency from unity are all the mentioned actions of generalized friction, thermal, chemical, mechanical and electrical.

The particular expressions of the efficiency depend upon the aim of the object.

If the power plant is considered, giving power only:

$$\eta_\varepsilon = \dot{\varepsilon}^{w}_{out} / \dot{\varepsilon}^{m}_{in} \qquad (3.3)$$

For the cogeneration plant it should be

$$\eta_\varepsilon = \frac{\dot{\varepsilon}^{w}_{out} + \dot{\varepsilon}^{Q}_{out}}{\dot{\varepsilon}^{m}_{in}} \qquad (3.4)$$

It is the same as Fux formula (2.8) proposed in 1940. Here in the numerator the two exergy flows are summed. If to sum the energy flow up (power and heat) the misleading results appear.

For any boiler

$$\eta_\varepsilon = \dot{\varepsilon}^{Q}_{out} / \dot{\varepsilon}^{m}_{in} \qquad (3.5)$$

The same is valid for all the heat supply systems, like heat pumps – power plant system, where the compressor is fed from power plant and consumed fuel.

For electrical melting or thermal treatment:

$$\eta_\varepsilon = \dot{\varepsilon}^{m}_{out} / \dot{\varepsilon}^{w}_{in} \qquad (3.6)$$

For a chemical or metallurgical plant, which consumes all three kinds of exergy:

$$\eta_\varepsilon = \dot{\varepsilon}^{m}_{out} / (\dot{\varepsilon}^{m} + \dot{\varepsilon}^{Q} + \dot{\varepsilon}^{w})_{in} \qquad (3.7)$$

However for vigorous comparison, from the experience of the present writer, the reduction of $\dot{\varepsilon}^{Q}$ and $\dot{\varepsilon}^{w}$ to primary fuel consumption in assisting boiler houses and power plants should be recommended.

There exists a difference between exergy efficiency of a turbine and its internal (isentropic) one. The turbine power is $\dot{\varepsilon}^{w} = h_1 - h_2$=enthalpy drop. The exergy difference $\Delta\varepsilon = h_1 - h_2 - T_o(S_1 - S_2)$. Hence

$$\eta_\varepsilon = \frac{\Delta h}{\Delta \varepsilon} = \frac{h_1 - h_2}{h_1 - h_2 - T_o(S_1 - S_2)} = 1 + \frac{T_o(S_1 - S_2)}{h_1 - h_2 - T_o(S_1 - S_2)} \tag{3.8}$$

In contrast to it the isentropic efficiency

$$\eta_\varepsilon = \frac{\Delta h}{\Delta h_s} = \frac{h_1 - h_2}{h_1 - h_{2s}} = \frac{h_1 - h_2}{h_1 - h_2 - T_{av}(S_1 - S_2)} \tag{3.9}$$

here h_{2s}=enthalpy after isentropic expansion in a turbine, $T_{av}=(h_1-h_{2s})/(S_1-S_2)$=averaged temperature of expansion. These formulas (3.8) and (3.9) coincide only if $T_o=T_{av}$, which cannot happen in practice. In the gas turbine $T_{av}>T_o$, hence $\eta_\varepsilon>\eta_s$, however in the detander of a cooling machine $T_{av}<T_o$ and $\eta_\varepsilon<\eta_s$.

In MHD generators, where T_{av} is especially high, the exergy efficiency might be high even by small η_s.

For the heat transfer the efficiency might be expressed as a product of the two Carnot factors: for a reversed cycle with higher temperature T_2 and a direct power cycle with lower one T_1:

$$\eta_\varepsilon = \frac{T_1 - T_o}{T_1} \cdot \frac{T_2}{T_2 - T_o} \tag{3.10}$$

The less the difference of T_2-T_1 the closer η_ε is to unity. If T_2=2000K, T_1=800K, T_o=300K the η_ε=0,73.

For heat engine, either turbine or piston, where the heat flow Q is consumed by temperature T and a mechanical power $N = \dot{\varepsilon}^W$ is produced we have:

$$\eta_\varepsilon = \dot{\varepsilon}^w / \dot{Q}(1 - \frac{T_o}{T}) \tag{3.11}$$

For a heat pump the reversed formula is valid. Here $\dot{Q}$ is the useful heat and $\dot{\varepsilon}^w$ is the consumed electrical or mechanical power.

If we deal with the combined machine (heat transformer) or consider the total fuel-fired power plant system and heat pump the ratio of delivered and consumed heat is used; coefficient of performance (COP):

$$COP = \overset{e}{\eta_\varepsilon} \cdot \overset{hp}{\eta_\varepsilon} \frac{T_2 - T_o}{T_2} \cdot \frac{T_1}{T_1 - T_o} \tag{3.12}$$

in contrast to (3.10) here the Carnot factor for direct cycle and reversed Carnot factor for inversed cycle are used.

In the ideal case, when $\eta_{\varepsilon}^{e}=\eta_{\varepsilon}^{hp}=1$, T_2=800K, T_1=400K, T_o=300K we have COP=2,5. It means that the ideal heat transformer in comparison to ideal boiler house (η=1) can give 2,5 times more heat with the same fuel consumption or consume only 40% of fuel, giving the same useful heat.

Here there is no violation of the First Law, because of the income of low-grade heat flow from the environment (atmosphere or river) to the evaporator of a heat pump.

In the case of reasonable partial efficiency of the engine and heat pump $\eta_{\varepsilon}^{e}=\eta_{\varepsilon}^{hp}$=0,7 the COP would be 1,25. It gives nearly a two-fold fuel economy in comparison to real a small boiler with efficiency $\eta_1 \approx 0{,}65$. If we could use the rejected heat from the engine, the COP might increase to 1,5-1,6.

This approach is the ideal way to save fuel, provided the consumer could use heat with the temperature deflection from the ambient one on 50-60°C, not more. It is the large area of heat supply systems for domestic heating.

The electrical heater presents an opposite example, exergy efficiency of such a space heater is nearly 0,05 due to large exergy loss from electrical resistor to room air. It is ten times more energy expensive than thermal transformers.

Energy and Exergy Efficiency of Cogeneration and Boiler Houses

Let us consider gradually the complicated scheme of combined heat and power production, referred to as cogeneration (Fig. 3.1)

The consideration consists of three steps.

I. At first it is assumed as a boiler only, aimed to heat water in a heating system.

II. The gas turbine is installed before the boiler, driving a generator to produce electricity. The exhausts, after the turbine, give some heat in heat exchanger.

III. The steam turbine plant is installed in between the gas turbine and water heater, which serves as a steam condenser. This is the full scheme of the combined gas-steam cogeneration.

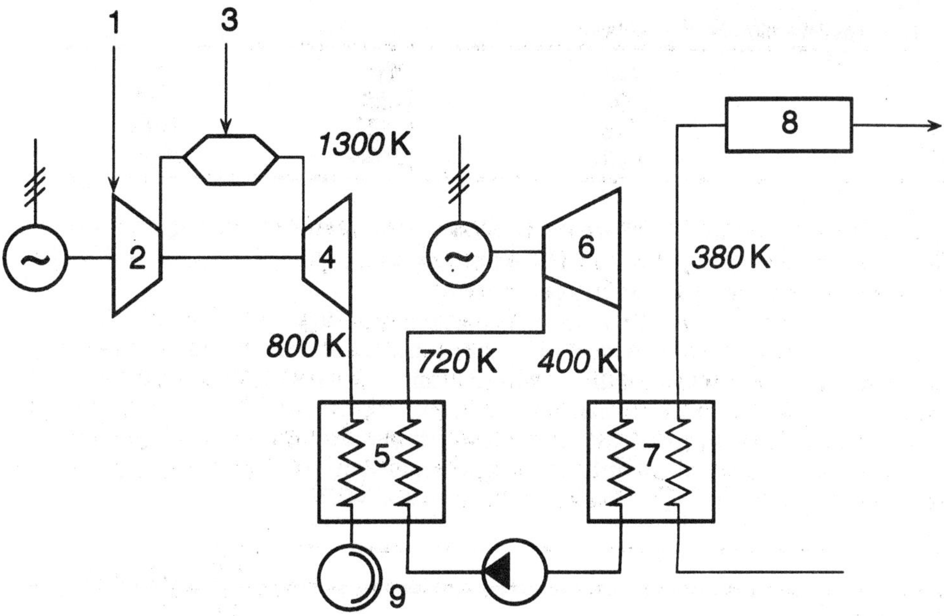

Figure 3.1 Outline of combined gas–steam cogeneration plants.

1 - air
2 - compressor
3 - combustion chamber
4 - turbine
5 - heat exchanger
6 - steam turbine
7 - water heater
8 - peaking boiler
9 - stack

Now we will calculate, looking at the temperatures given in Fig. 3.1, the three efficiencies: electrical η_{el} First Law η_1 and exergy η_ε. The reference temperature T_o=270K.

I. Here $\eta_{el}=0$; $\eta_1=0{,}85$; $\eta_\varepsilon=0{,}85\ (380-270)/380=0{,}25$.

II. If $\eta_{el}=0{,}28$ and for heat exchanger $\eta_1=0{,}65$ we get:

$$\eta_1=0{,}28+(1-0.28)\ 0.65=0.75$$

$$\eta_\varepsilon=0{,}28+(1-0.28)\cdot 0{,}65\cdot (380-270)/380=0{,}416$$

III. If η_ε for the steam cycle alone is assumed 0.6, the $\eta_1=0.6\ (720-400)/720=0.265$. Therefore the total plant:

$\eta_1=0.28+0.72\cdot 0.65\cdot 0.265+0.72\cdot 0{,}735\cdot 0{,}65=0.744.$

$\eta_\varepsilon=0.28+0.124+0.34(380-270)/380=0.50.$

The results are in the table:

	η_{el}	η_1	η_ε
I.	0	0.85	0.25
II.	0.28	0.75	0.416
III.	0.404	0.744	0.50

Here, one can see that the more complicated and useful the plant, the less its First Law efficiency η_1. It is an inherent illness of η_1. On the contrary, the exergy efficiency reflects usefulness correctly.

Let us consider now the comprehensive compilation of Horlock Fig. 3.2. Here we see ten schemes of energy objects, from simple boilers to combined gas-steam turbine power plants. At the entrance, the fuel inflow was assumed F=1. At the exit, in first and second columns, the power W and heat Qu, as the part of F=1, were presented. For these figures the author is responsible.

For exergy minded people it is quite evident that this comparison is misleading for a boiler Qu=0,9 and gas turbine W=0,3.

<table>
<tr><th></th><th></th><th>W</th><th>Q_U</th><th>Z_2</th></tr>
<tr><td>A. Domestic Boiler</td><td>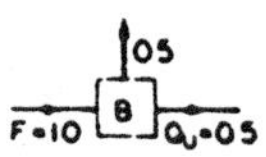
</td><td>0</td><td>0.5</td><td>0.135</td></tr>
<tr><td>B. District heating (industrial) boiler</td><td>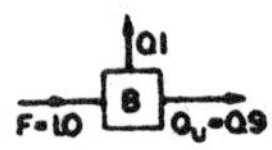
</td><td>0</td><td>0.9</td><td>0.24</td></tr>
</table>

(Figure 3.2 continued on the following two pages)

Figure 3.2 Exergy efficiency for ten energy schemes

C. Gas turbine power plant	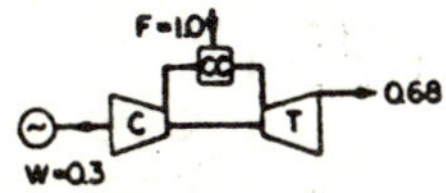	0.3	0	0.30
D. Steam turbine power plant	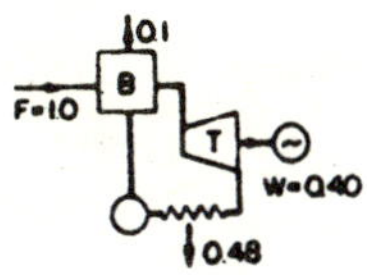	0.4	0	0.40
E. Combined gas turbine/steam turbine power plant	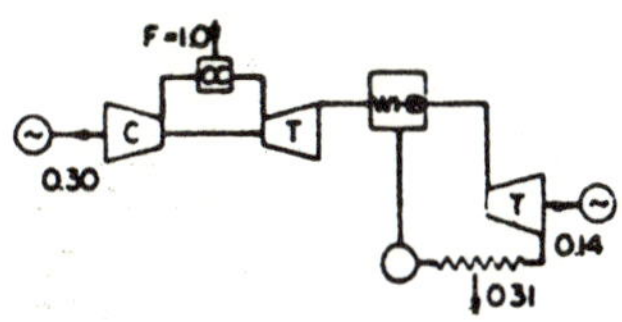	0.44	0	0.44
F. Extraction condensing plant	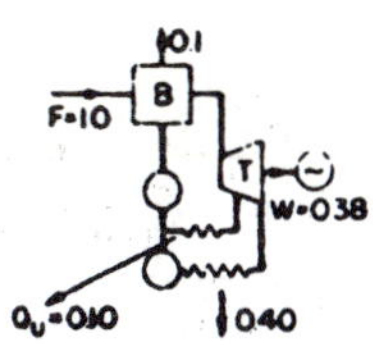	0.38	0.10	0.41

(Figure 3.2 continued)

G. Back pressure plant	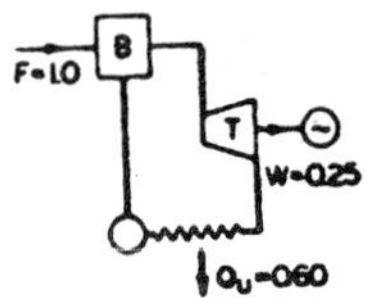	0.25	0.60	0.41
H. Gas turbine with waste heat recuperator (WHR)	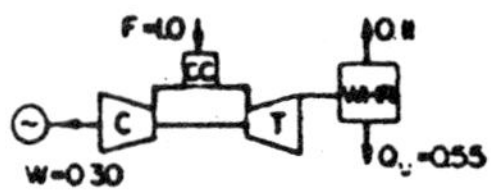	0.30	0.55	0.45
J. Combined cycle (gas turbine/back pressure steam turbine	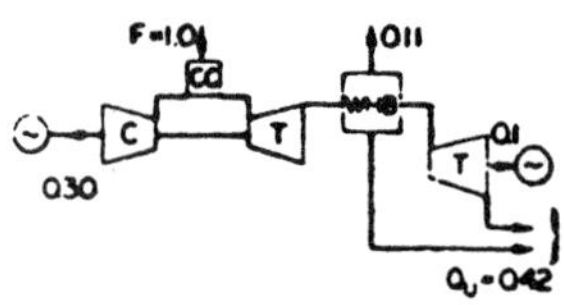	0.4	0.42	0.51
K. Heat pump driven by back pressure steam turbine (heat only scheme)	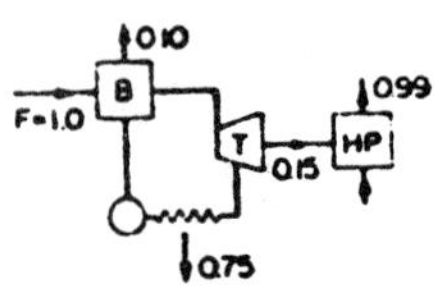	0	1.74	0.47

(Figure 3.2 continued)

The exergy efficiency is here as follows:

$$\eta_2 = W + \frac{\Delta T}{T} Qu = W + 0,27Qu$$

The temperature of the heat supply was assumed equal (100°C) and the reference temperature 0°C, $\Delta T/T=100/373=0,27$.

The gradual increase of efficiencies seems to be reasonable; it perfectly reflects complexity of the devices, diminishing exergy losses. Some figures here differ from the previous example, but the general picture is the same. The exergy efficiency only seems to be of value.

Exergy Allocation of Fuel Consumption in a Cogeneration Plant

There exist many different ways to evaluate specific fuel consumption and allocate consumed fuel over the power and heat. Here the two: Distraction method and Exergy one are to be considered.

In Distraction method a cogen plant is considered as replacing an ordinary boiler house. Therefore the logic of the method is to distract the fuel consumption of the replaced boiler house B_b from the fuel flowrate to cogen plant Bt and divide by produced electric power N:

$$b = \frac{B_t - B_b}{N} \tag{3.13}$$

In exergy methods the flows of exergy are calculated and the income flow is divided by the flowrate of delivered exergy:

$$b = \frac{B_t}{N + \tau Q} \tag{3.14}$$

Here $\tau = \frac{T - T_0}{T}$-exergy temperature function (Carnot factor), Q-heat flow, T_0-reference temperature.

For the sake of simplicity all the flows are considered in energy units, joules or watts and some dimensionless variables are introduced:

$\eta=N/B_t$-electrical efficiency of cogen plant,

$\alpha=(N+Q)/B_t$-share of used energy on cogen plant,

$\beta=Q/B_b$-share of used energy in the boiler house.

The evident inequalities are : $\alpha<1$; $\beta<1$; $\eta<\alpha$.

After some simple algebra one can get:

$$b = \frac{B_t - B_b}{N} = \frac{B_t - \frac{1}{\beta}(\alpha B_t - \eta B_t)}{\eta B_t} = \tag{3.15}$$
$$= \frac{1}{\eta}\left(1 - \frac{\alpha}{\beta}\right) + \frac{1}{\beta} \qquad \textit{(Distraction)}$$

$$b = \frac{B_t}{N + \tau Q} = \frac{B_t}{\eta B_t + \tau(\alpha B_t - \eta B_t)} = \tag{3.16}$$
$$= \frac{1}{\eta + \tau(\alpha - \eta)} \qquad \textit{(Exergy)}$$

Having looked at (3.15) one can see the strong dependence of b upon the relation α/β which is very close to unity. If $\alpha=\beta$ the b does not depend on quality of power conversion process, it is simply energy efficiency of replaced boiler in minus one degree $1/\beta$.

If $\alpha>\beta$ the first term is negative. The better is power plant (the higher η), the greater is specific fuel consumption. Arbitrary value of b might be obtained by proper selection of small deviations of α or β, it is evident on Fig. 3.3. In case $\alpha<\beta$ a very small decrease of b is seen in spite of large increase of η.

All these misleading facts let us consider this approach as absolutely false.

Exergy method calculations are presented on Fig. 3.4. Here one can see gradual decrease of b with increase of η which stimulate the use of good machines and cycles. Allocation of consumed fuel in proportion to exergy flows seems to be quite natural. Separating line equation is as follows:

$$b_{el} = \frac{\eta}{[\eta + \tau(\alpha - \eta)]^2} \tag{3.17}$$

Below this line is fuel consumption for electricity, above it the shaded area reflects the same for heat supply.

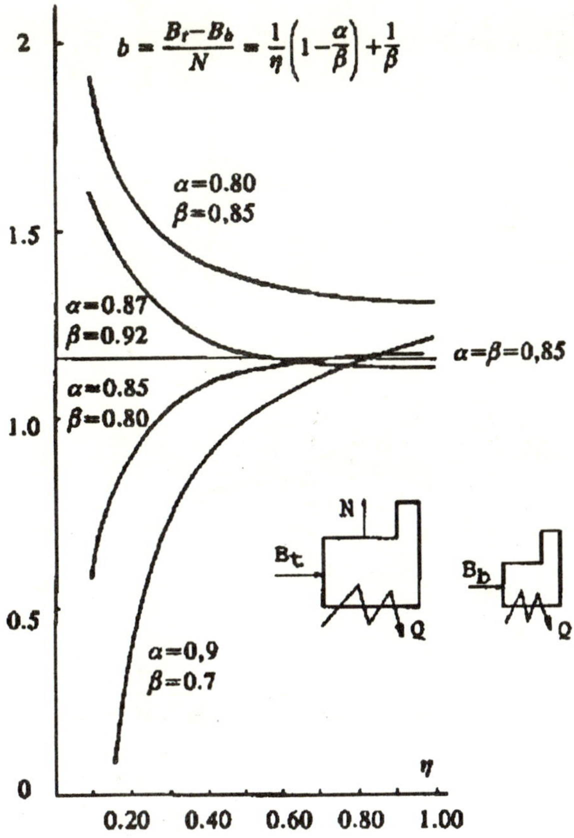

Figure 3.3 Specific fuel consumption versus electrical efficiency according to distraction method.

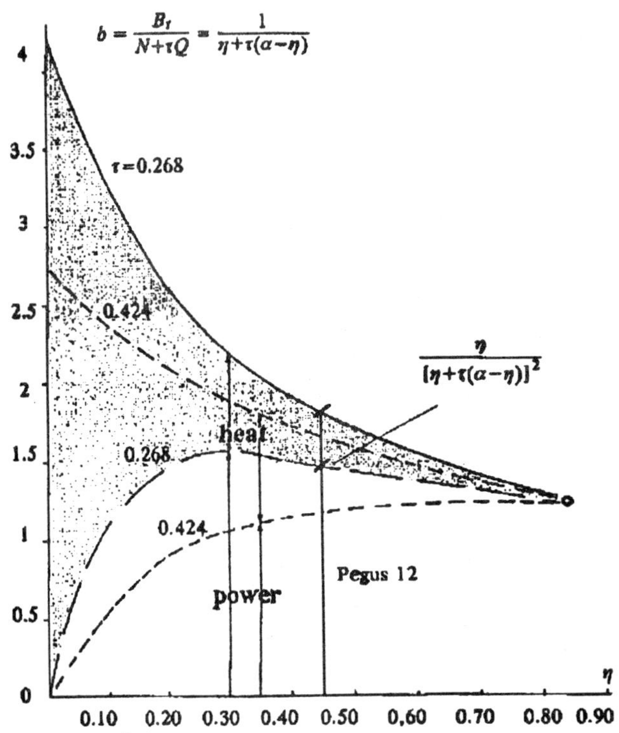

Figure 3.4 Specific fuel consumption based on exergy method (T_0=0°C, α=0,85)

The value of b when η=0 presents a boiler house. The higher η the better is our cogen plant, which describes advanced and combined cycles. Two temperatures of delivered heat 100C (τ=0.268) and 200 C (τ=0.424) reflect district heating and process steam respectively. The higher is temperature the less is specific fuel consumption. Exergy approach seems to be valid for any cogeneration plant.

Let us consider as an example the best gas-fired cogen plant Merwedekanaal, city of Utrecht, PEGUS-12 unit. It consists of the gas turbine ABB type 13E, steam bottoming turbine and heat supply system for municipal service.

According to measurements (Ziegler, et al.) here N=209.1 MW, Q=182.7 MW, cooling water temperature t=4°C and district water t=105°C. In this regime electrical efficiency η=0.45, α=0.85 and τ=(105-4)/278=0.268.

Specific fuel consumption

$$b=[0.45+0.268(0.85-0.45)]^{-1}=1.80$$

per unit of delivered exergy. The fraction of fuel consumed for power equals

$$b_{el}=0.45[0.45+0.268(0.85-0.45)]^{-2}=1.45.$$

The rest 1.80-1.45=0.35 is the fuel exergy consumption for heat supply. Note here the neglection of small difference between fuel (natural gas) exergy and combustion heat. See this distribution on Fig. 3.4.

In percentages we see 80.5% fuel is consumed for power and 19.5% for heat.

If the gas turbine only is on duty (145 MW) and steam turbine is shut down in favor of district heating, the same values would be as follows: $\eta=0.31$, $b=2.20$, $b_{el}=1.50$, the rest for heat 0.70, in percentages 68% for power and 32% for heat.

Specific fuel consumption is increased due to exergy losses in large temperature drop of water heaters.

Comparisons of Cumulative Exergy Efficiencies

It is well known that in exergy consumption calculations, the reference state and the particular technological process must be specified. Many opponents of exergy calculations used to mention the variety of such processes. We would like to present the results of calculations made by two research groups, working on two diametrically opposite points of the globe without any links between them: the first group is headed by Szargut and Morris in Canada, and the second one by Stepanov and Stepanova in Irkutsk.

These two groups calculated the exergy efficiency (η_ε), i.e., the relation between the chemical exergy of the most widely used metals and the consumed exergy. The metals include steel and practically pure elements such as Al, Cu, Pb, Zn. The exergy consumption includes the exergy to get raw materials from ore in the ground and the fuel equivalent of the energy consumed. In Stepanov's work, only one process was taken into account. In Szargut-Morris' work, some variants were considered and the discrepancy is indicated in the graphs of Figure 3.5.

It is evident from the Figure that the correlation is good. This cannot be a coincidence since six different cases are compared. The only difference, in the case of electrosteel production, might be explained by the way the primary energy is embedded to the scrap that it is accounted for. The comparison of the results in Figure 3.5 show that exergy calculations reflect facts independent of the particular investigators.

By the way, in view of the exergy efficiency, one can see in Fig. 3.5 the extremely low figure for copper and lead. This cannot be inherent. In the technology associated with these metals, a large energy conservation should be expected.

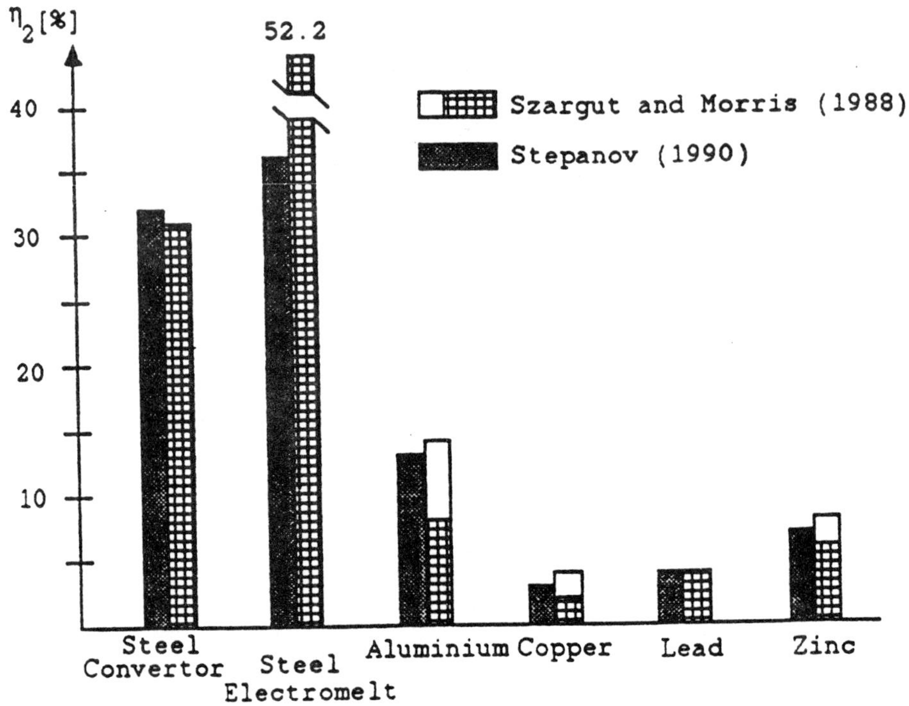

Figure 3.5 Comparison of the exergy efficiency (η_ε) values of several metals calculated by Szargut and Morris, and Stepanov.

Chapter 4.

EXERGONOMICS

Saving is a very good thing. Especially when your parents have done it for you.
Winston Churchill (Attrib.)

Net-exergy

The word "net-exergy" means an important thing in energy supply calculations. To explain it, we will at first specify the "exergy-brutto." It is not a flow of exergy, but the total amount of delivered exergy by energy supplier (boiler house, power plant etc.). If we take into account that this supplier appeared not by magic, we will evaluate the amount of exergy which has been spent to create it. The net-exergy means the total delivered exergy less the consumption to build the object and produce equipment.

To this calculation we have to generalize the exergy balance, with respect to flows, crossing the considered object boundary. Before the construction inside this boundary was essentially nothing, except total equilibrium of atmosphere and ore in the ground.

During the construction procedure, the boundary has been crossed by flow of exergy in materials and fuels. The history of any object begins from the spending of exergy, not of delivering.

After some years of construction work, as well as equipment creation (which means the consumption of exergy) the makings of the object has begun. After some time of operation, it returns the consumed exergy in the form of electricity or heat. After the lifetime the operation is cancelled. All these events are presented on the graphs Fig. 4.1 and 4.2. In the Fig. 4.1, the exergy flow input $\dot{\varepsilon}_{in}$, along with assist exergy flow $\dot{\varepsilon}_{assist}$ is presented. The useful delivered flow is $\dot{\varepsilon}_{out}$. The formerly consumed exergy to build the object is ε_b; it is the area over the left curve on Fig. 4.2.

The equal area behind the curve on the right of the vertical axis determines the exergy payback time t_{pb}. The rest of the gained exergy is the net-exergy until the end of the exploitation on the time τ.

Here appears an important coefficient of net-exergy which has been calculated for many energy objects:

$$K_\varepsilon = \frac{\dot{\varepsilon}_{out} \cdot \tau}{\varepsilon_b} \tag{4.1}$$

The typical figures for fossil fuel fired power plants are 4-7, for natural gas near the well it is nearly 100. However, at the far located consumer it would be ten times less. (Cleveland, 1984; Critopf, 1982; Tsoulfanidis, 1981; Gilliland et al.; Herendeen, et al.)

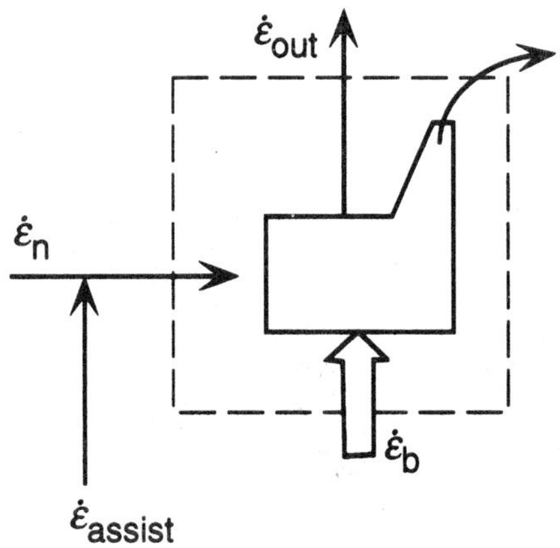

Figure 4.1 Exergy flows through boundary.

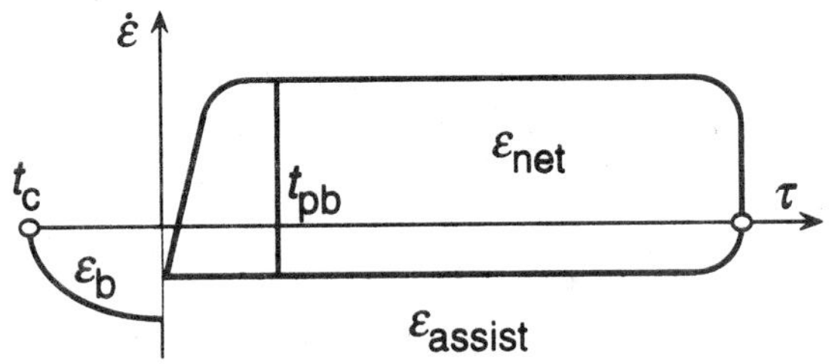

Figure 4.2 The graphical history of an energy object.

Exergy Price List for Construction Materials

The main part of exergy needed to build an object, consists of material exergy to melt the metals to form resins of oil etc. In many machines this share is up to 90%. The turning, welding, assembly need not be more than 10-20%. Hence, to calculate exergy consumption for construction, the most important data are energy intensities (specific consumption per kg) of used materials.

Hitherto a unique collection of these data is the book of Boustead and Hancock (1979). Then , J. Szargut et al, have reexamined that handbook in order to reflect the exergy approach. Many data are presented by V. Stepanov in his tabbs on exergy of elements and chemical compounds.

Table 4.1 Specific exergy intensity of some substances tce/t

Name	Sign	Boustead	Kellog	Szargut	Stepanov
Titanium	Ti	34	15.1		
Mercury	Hg		14.7		
Magnesium	Mg	16	13.1		
Tungsten	W		12.9		
Nickel	Ni	24	5.3	8	8.5
Molybdenum Oxide	MoO_2		5.5		
Aluminum	Al	8.4	9.0	9.8	8.1
Vanadium	V			11.6	
Vanadium Oxide	V_2O_5			12.5	
Silicon	Si	7.5			
Phosphorus	P		6.33		
Copper	Cu	4.3	4.13	4.03.9	
Natrium	Na	5.0	3.4		
Zinc	Zn	3...6.2	2.4	2.6	
Steel (rolled stock)	Fe	1.8..2		1.6...2	
Steel	Fe	1...1.5	0.9	0.7...1	
Magnesium Oxide	MgO	1.0			
Sulphur	S	0.2			
Arsenic	As		0.1		
Lead	Pb	1.7			1.4
Cement		0.17			

We restrict ourselves the comparable table 4.1 with exergy intensity of some important materials expressed in tce/t. As to exergy efficiency of material production, the big list has been collected by Szargut et al (1988), see also Fig. 3.5. The discrepancy in figures could be explained by the difference in technology used (see also Kellog).

Main Criterion of the Exergonomics

Having looked at Fig. 4.1 and 4.2 we can calculate the total exergy efficiency of any energy object. The word "total" reflects our ability to take into account the formerly consumed exergy ε_b along with the current expenditure.

The delivered exergy over the total life time τ is $\dot{\varepsilon}_{out} \cdot \tau$, the total consumption equals $\dot{\varepsilon}_{in} \cdot \tau + \varepsilon_b$ hence

$$\eta_{tot} = \dot{\varepsilon}_{out} \cdot \tau / (\dot{\varepsilon}_{in} \cdot \tau + \varepsilon_b) = \left(\frac{\dot{\varepsilon}_{in}}{\dot{\varepsilon}_{out}} + \frac{\dot{\varepsilon}_b}{\dot{\varepsilon}_{out} \cdot \tau} \right)^{-1} = 1 / \left(\frac{1}{\eta_\varepsilon} + \frac{1}{K_\varepsilon} \right) \tag{4.2}$$

All the values in the rhs are known already. We can see, that $\eta_{tot} < \eta_\varepsilon$, the limiting case $\eta_{tot} \to \eta_\varepsilon$ when $K_\varepsilon \to \infty$.

The denominator of η_{tot} (or inversed quantity) is

$$Z = \frac{1}{\eta_\varepsilon} + \frac{1}{K_\varepsilon} \tag{4.3}$$

it is just the sum of specific exergy consumption (SUSEX), the expenditure per unit of actually delivered exergy, the main criterion of the exergonomics. It is similar to a monetary cost of energy, where monetary cost is replaced by exergy one. It may play the role of a functional to be minimized in optimization problems.

The simplest way to minimize Z we can find when the function $K_\varepsilon(\eta_\varepsilon)$ is known. As usual, both the K_ε and η_ε are known functions of energy current density δ. Increasing δ leads to increase of K_ε and decrease of η_ε. Once $K_\varepsilon(\eta_\varepsilon)$ is given, the simple equation $\frac{\partial Z}{\partial \eta_\varepsilon} = 0$ roots are

$$\eta_\varepsilon^{opt} = K_\varepsilon / \sqrt{-\frac{dK_\varepsilon}{d\eta_\varepsilon}}; \; Z_{min} = (1 + \sqrt{-\frac{dK_\varepsilon}{d\eta_\varepsilon}}) / K_\varepsilon \tag{4.4}$$

The increase of η_ε over the η_ε^{opt} leads to the wasteful exergy consumption.

The real measure in energy economy is $M=(Z^{real} - Z^{min})/Z^{real}$. The coefficient of utility in exergy consumption is given by $1-M=Z_{min}/Z^{real}$.

As the model function for illustration we can take $K_\varepsilon(\eta_\varepsilon)$ relation in the form $K_\varepsilon = A\eta_\varepsilon^{-m}$, where A and m are constants, $0<A<m$.

In this case

$$Z_{min} = \left(\frac{m}{A}\right)^{\frac{1}{m+1}} + \frac{1}{A}\left(\frac{A}{m}\right)^{\frac{m}{m+1}}$$

If A=0.5 and m=2 the η_ε^{opt}=0.63, Z_{min}=2.37.

To reflect the exergy discounting we include parameter $\lambda=\tau_\eta/\tau$, the ratio of a given normative time to the total life time. Now $Z=1/\eta_\varepsilon+1/K_\varepsilon\lambda$. The variation of Z versus η_ε by different exergy discount rate λ is presented in Figure 4.3. The existed minimum of Z versus η_ε is evident, the less λ the less the η_ε^{opt}. It is a general behavior of Z criterion. The unity of η_ε is impossible to achieve due to infinite exergy consumption ($Z\rightarrow\infty$). In effect it is a reformulation of the Second Law. The very interesting statistics is presented on the graphs of Figure 4.4 (Haefele). The history of three different objects is compared by means of exergy efficiency rise over the years. One thing is evident, the halt of increase after 35-40%. One can wonder, why there is so much room to rise. The possible answer could be found on previous Figure 4.3: it is just optimal exergy efficiency by the λ=0.25.

Some early ideas on the matter one can find in deVries and Nieuwlaar, Hannon papers.

The SUSEX formula (4.3) appeared first in the book of Yantovskii and Poustovalov in 1982.

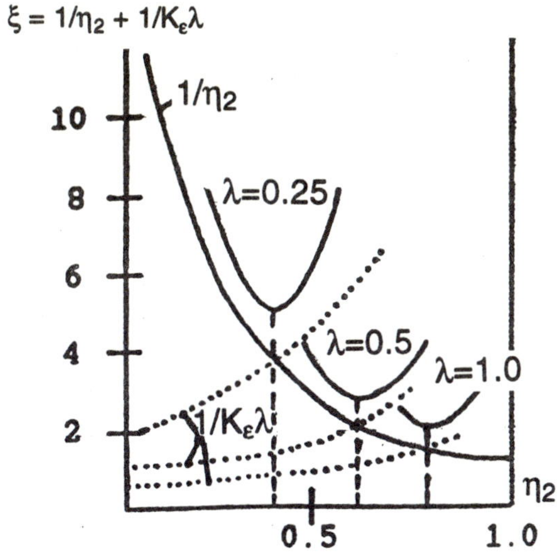

Figure 4.3 Variation of the exergy expenditure $Z(\eta_\varepsilon)$. (SUSEX)

Energy Supply Complex as a Conductor of Exergy

Let us compare the important cases of exergy accounting.

The two flows which are being compared, strongly differ by their complexity:

- electrical energy flow in a direct current conductor, where the only exergy carrier is an electrical charge.
- exergy flow in energy supply complex of a country, where at least four carriers (charge, substance, impulse and entropy) are working.

It is unnecessary to say how complicated the latter is to the former. But in principle there are no differences if we take into account, in the second case, all the kinds of generalized friction; not only electrical friction of electron gas in the conductor (Ohm's resistance) but also chemical friction (mixing), mechanical friction in bearings and, the most important, thermal friction in a heat transfer by large temperature drop.

This analogy seems to be a useful tool in a logic of computerization of any energy spiders and in energy engineering education as well. Let us consider the simplest energy transfer device, the insulated conductor. Its length L, cross-section F, current J, voltage V and total energy intensity of metal production Z are known.

Neglecting current leakage through insulation, the total energy expenditure is as follows:

$$D=J^2L\tau/\sigma F\eta + Z\rho LF \tag{4.5}$$

Here the first term is Joule heating in a time period τ, converted to primary energy (fuel) by means of efficiency η.

The exergy efficiency of the conductor itself coincides with the ordinary electric one:

$$\eta_2 = JV/(JV+J^2R) = 1/(1+\frac{j_q L}{\sigma V}) \tag{4.6}$$

For an ordinary electric line the formula should read:

$$j_q=10^6 A/m^2,\ L=10^6 m,\ \sigma=2\cdot 10^7 Sm/m,\ V=10^5 V,\ \eta_2=0{,}9$$

As the cross-section F in (4.5) increases so does then the first term decrease and the second one increase. The optimum conditions are reached when the two terms in (4.5) become equal (Kelvin Law).

The optimal energy carrier density (current density) is easily obtained by means of the equation $\frac{\partial D}{\partial F}=0$

$$j_q^{opt} = \sqrt{Z\rho\sigma\eta/\tau} \tag{4.7}$$

Let be for illustration η=0,35, Z=330 MJ/kg, for aluminum, $\tau=4 \cdot 10^8$ sec, then j_q^{opt}=0,12 A/mm^2, it's ten times less than in common practice.

Comparable Items

Electrical conductor	Energy supply complex
Electrical insulationto prevent current leakage	Pipe walls, heat insulation, electrical one
Power loss due to Ohm resistance	Exergy loss due to generalized friction
Entropy current to environment, carrying Joule heat	Entropy current to environment, carrying exergy losses
Exergy to create conductor (metal, resin)	Exergy to produce all the equipment
Optimization of electric line by the sum of Joule losses fuel consumption and fuel consumed to build conductor.	Optimization of the complex by the sum of exergy current losses and exergy consumed to build equipment.

Now we can put down the formula for the sum of losses, analogous to (4.5) for conductor:

$$D = \sum T_{oi} \Delta Si + \dot{\varepsilon}_v + \dot{\varepsilon}_{eq} \tag{4.8}$$

where the first term of rhs is the sum of lost works, then is the external exergy loss and the last exergy to build equipment.

In the simplified case of steady-state energy system, without an increase of the total primary energy consumption, the situation is similar to the one mentioned above. Like in the simple conductor, we may write down the general formula for total exergy loss:

$$\dot{D} = T_o \sum \Delta \dot{S} + \dot{E}_{equip} \tag{4.9}$$

Here $\sum \Delta \dot{S}$represents the sum of entropy increase in all parts of the system and $\dot{E}_{equip}$ is the exergy expenditure to manufacture the equipment and all the auxiliary devices of the energy system. The total exergy efficiency of national energy system is

$$\eta_\varepsilon = \frac{\dot{E}_{deliv} - \dot{E}_{equip}}{\dot{E}_{extract} - \dot{E}_{export}} \tag{4.10}$$

The indexes explain the meaning of the variables in the equation (4.10). In the numerator, we put the delivered exergy minus the exergy consumed by the energy system itself: in the denominator we have extracted exergy minus the exported one, if any.

For example, the government of a country launched a bill to decrease the exergy loss in boilers by means of increasing boiler surface and the decreasing of temperature drop. In the same time the amount of fuel used in metallurgy plants will increase and the real fuel savings in that country are questionable.

An example of the rough evaluation of the amount of fuel savings in the heat supply system of the USSR is presented using the assumption that exergy efficiency of a heat supply system (0.24), in the year 2000, will approach the value of 0.34. This is reasonable because it is less than the well known limit in Fig. 4.3 and the practical means (cogeneration, cascading, heat pumps) have already been developed. The estimated fuel savings was 220 mtce. This figure is worth the efforts required to achieve it. The exergy flow diagram of the USSR energy complex, see on Fig. 4.5. It is a unique diagram, because before there were no such practice and after the object of analysis disappeared (USSR)

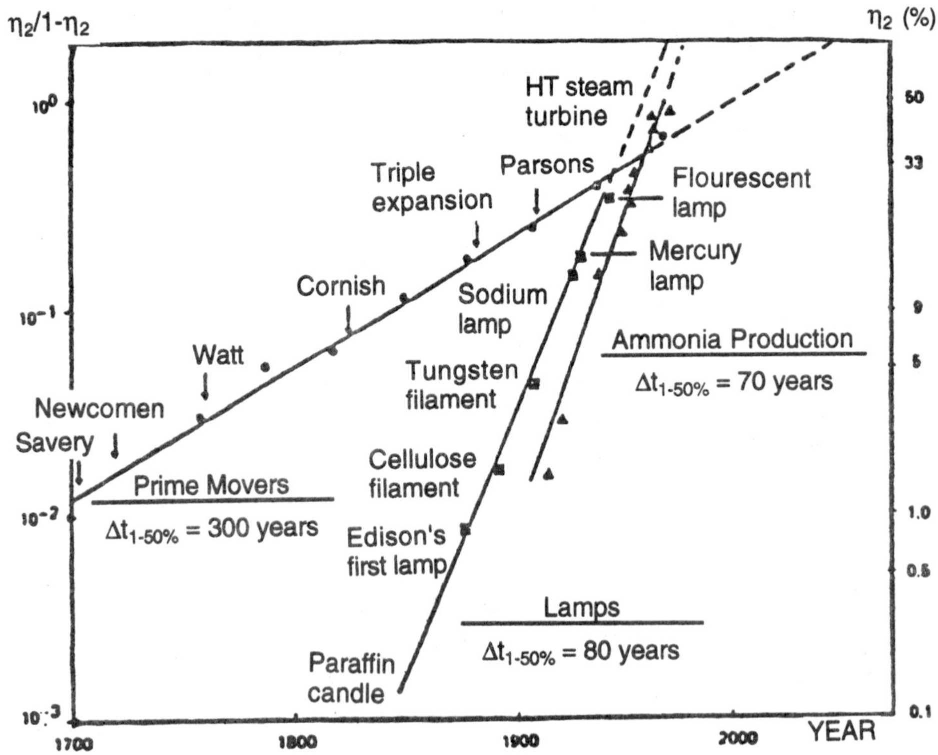

Figure 4.4 Historical trends in second law efficiencies (η_2). $\Delta t_{1\text{-}50\%}$ is the time necessary to evolve from an efficiency of 1% to one of 50%.

The sum of the specific exergy expenditure ε_E for the whole energy complex of a country seems to be the invert variable of η_ε, equation (4.10): $\varepsilon_E=1/\eta_\varepsilon$. Therefore, a generalization of exergy analysis for a complex steady-state energy system seems to be feasible.

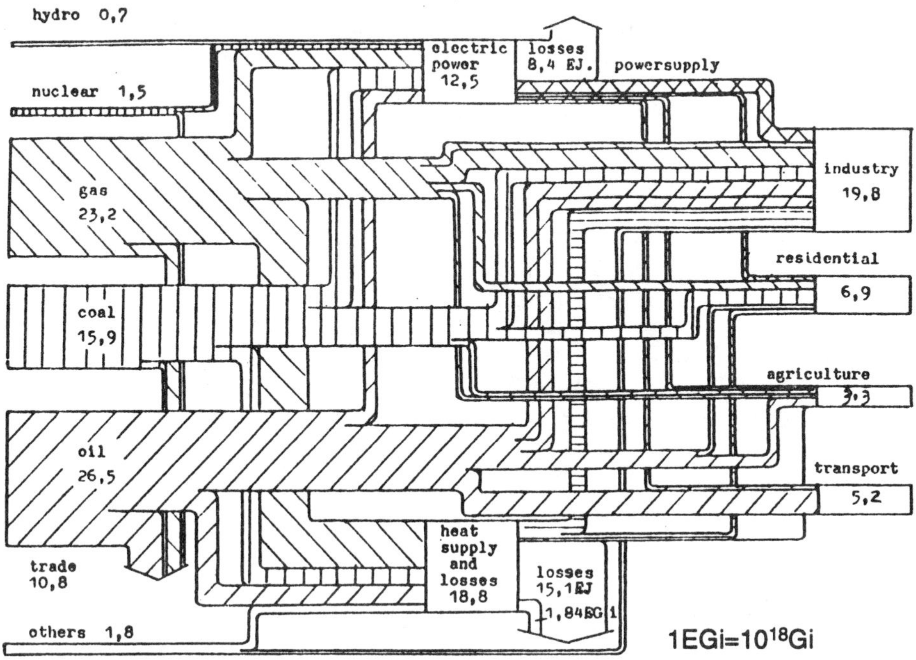

Figure 4.5 Exergy current diagram for the total energy supply complex of the USSR in 1985. 1Gi=1J of exergy.

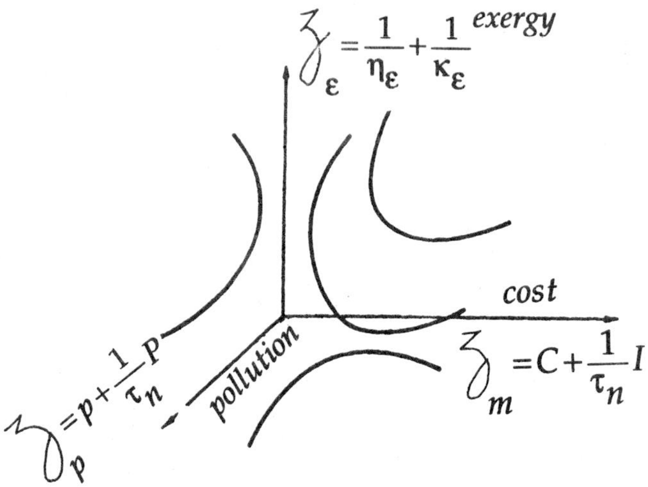

Figure 4.6 The characteristic curve in three-dimensional space money-exergy-pollution (risk).

Three-dimensional Space of Social Will

We have seen the important role of exergy currents in many problems. Certainly, exergy itself isn't sufficient to decision making for energy issues (see Alessio). However, profit itself, from the point of view of society as a whole, isn't sufficient too. There are possibilities of conflicts between the society and private owners.

The very complicated thing like social will, as the modern history of many countries vividly shows, can not be measured by only one value, even such general as the profit really is. Nevertheless the need to measure, to calculate, or at least to present on explicit graphs, the generalized properties of energy objects still exists.

Many values might be taken into account to describe energy supply. As the present writer is thinking, the best are three: monetary cost of energy (COE), sum of specific exergy consumption (SUSEX), pollution or risk to mankind (PRM).

Some correlations exist between them. But numerical dependence is expressed in formulae hitherto unknown. There are many controversies, for example what is externality (part of COE, caused by fines and taxes for pollution). The short-term (tactical) problem here is the lack of one man, sufficiently skilled in economics (COE) thermodynamics (SUSEX) and social hygiene (PRM).

The palliative solution to this problem might be found if we consider all three values as quite independent and award each item by its own coordinate axis. Here we follow Rene Descartes, who had arithmetized the physical space three centuries ago. He introduced three rectangular axis which formed a frame and reference system, and attached to any point of the space the three numbers (coordinates).

If there exists any link between numbers, it represents a visual item, line, curve, plane, surface. They are the subjects of differential geometry. Neither analytical mechanics nor relativity theory could appear without the space arithmetization.

The social will is much a more complicated subject than empty space. Therefore, even the possibility to visualize what we want is a headway. Hence, the present writer's proposition consists of presentation of all three values in the rectangular coordinate system:

X (COE) - Y (SUSEX) - Z (PRM)

if we consider them as independent coordinates, some explicit curves might be depicted.

This proposition isn't new. In 1979, Willem van Gool and independent Pierr Le Goff proposed plane curves, exergy versus money. It was a further development of isoquanta lines, well known in mathematical economics. "There are no news under the heaven."

The general outlook on the proposed three-dimensional curve is presented in Fig. 4.6. It is quite an illustrative curve, there is no one for real task as yet.

Certainly it is difficult to make a decision looking at such line. Any decision maker wants to launch an optimal issue. Here well known Pareto optimization might be used, however it is beyond the scope of the present book.

There exists a similarity in the structure of the mentioned three values. The first part depends on current consumption or release, the second part depends of former ones. In monetary economics that are current expenditures and investment cost, in exergy analysis the current exergy consumption and exergy investment, measured by net-exergy coefficient. In the case of pollution these are the current release and former ones, when equipment has been manufactured.

The attempts of real analysis are known only on the coordinate plains of Fig. 4.6; it means the use of projection of the three-dimensional curve.

On the Fig. 4.7, the curves of exergy consumption versus monetary cost is presented. The curve reflects the choice of an optimal building wall thickness (alternating parameter the thermal resistance coefficient).

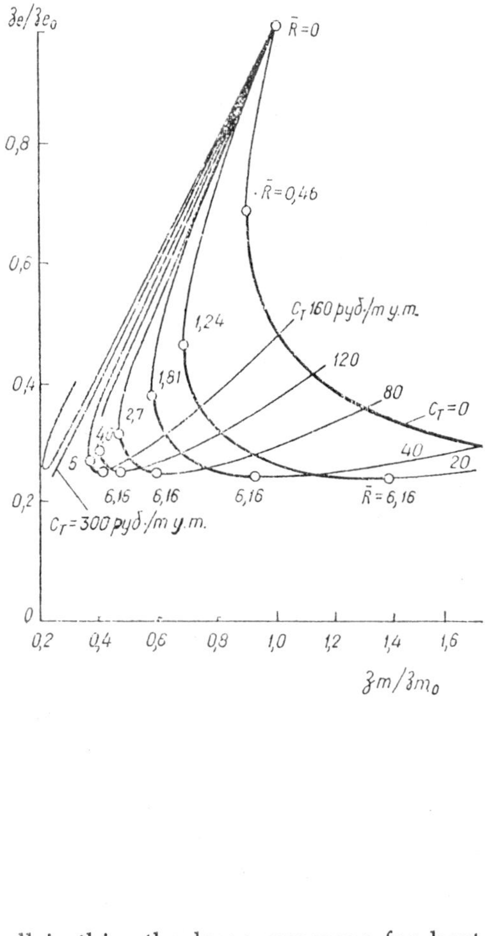

Figure 4.7

When the wall is thin, the large expenses for heat supply are needed, otherwise, for an extra thick wall the exergy and monetary spending for materials are increased. The series of curves reflects different fuel prices. (These plane curves were referred to as C-curves due to its shape.) In the Fig. 4.8, the plane curve for another horizontal projection was found in literature (Babaev, 1988). Here the risk versus monetary expenditure is presented. There are no links to pollution in vivid form, however, it is in hidden form. The coefficient of mortality is in close dependence of pollution.

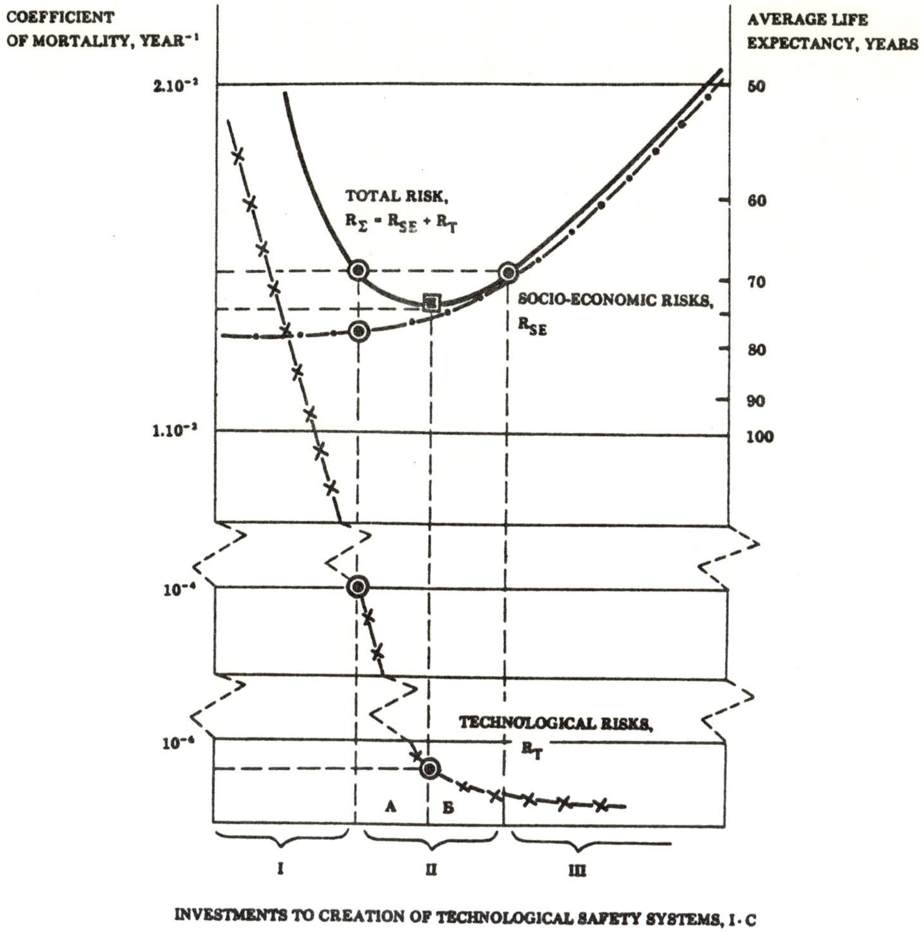

Figure 4.8 Optimization of expenditure to technological safety measures: I-the area of low expenditure leads to an unacceptable level of technological risk; II-the area of practically acceptable expenditure; III-the area of "unreasonably" high expenditure leads to unacceptable level socio-economic risk.

By the well known rules of geometry, having two projections, the third might be restored. Hence, the full three-dimensional curve might exist in reality.

Unfortunately no one real such curve is known to the present writer.

Computer Exergonomics of a Heat-Exchanger

All of the previous examples belong to "pre-computer" era and the calculations were simple and made by microcalculators at most.

The greatest advance could be made if a computer is used. In view of computerization, the special branch referred to as computer exergonomics might appear.

As an example, the C-curves made by means of modern computer procedure will be presented. The object of the calculations is the simple shell-and-tube heat exchanger. Here is this example:

It is necessary to warm initial water with cubic flow rate V2 from temperature T4 till T2. Initial temperature of hot medium (T1) and its expenditure (V1) may be changed in the given ranges in accordance with Fig. 4.9.

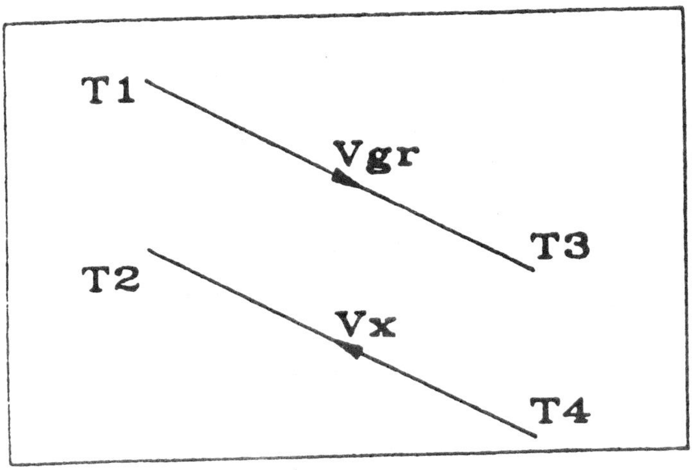

Figure 4.9 Diagram of temperature in the heat exchanger.

Task: It is required to select the optimum input temperature of hot medium (T1) and its expenditure (V1) and in accedence with them, the type and the number of sections of device, which satisfy conditions of minimization of criterion SUSEX and monetary expenses. The field of arc of choice on the "C-curve" between the two minimums, are relative to each axis.

Independent changeable variables in this task are input temperature of hot medium (T1) and its expenditure (V1) with limitations of the kind:

Tmin<=T1<=Tmax
Vmin<=V1<=Vmax

So, the final solution in the choice of required heat exchanger designer makes himself according to the calculated points of "arc of choice." That's why there's no necessity to hold global optimization. Let's restrict ourselves only with simple scanning of the field of permissible dimensions of changeable variables with given pace, successively calculating for every point corresponding criterion SUSEX and adduced expenditures according to which the known "C-curve" will be formed.

A mathematical model of heat exchanger, which is put in the program, represents the system of two equations: balance and transmission of heat. The solution of this system corresponds the stage of holding design calculation, in the logical scheme of the program, and is realized by method of sections. This method is applied in turn to every equation of the system with the check of errors of lowering of every variable to the root with the given errors. The cyclic stage of "sorting out standard line of heat exchangers by condition (adduced expenditures) AE→min", from our point of view, is more effective than choice of heat exchanger from standard line by condition of the optimum speed of the currents of media as it is recommended in many methodical manuals.

Exergy, which is used for pumping both carriers of heat (pure electric supply is fed to the drive of electropumps) is taken into consideration in calculations of criterion's SUSEX and adduced expenditures.

Exergy cost of creation of heat exchangers is calculated by corresponding data of the consumed material.

The program "SUSEX-exergy optimization of water-watery heat exchangers with making the C-curve" of choice is made in the medium Turbo Pascal (by Borland International, Inc.) and is realized on personal computers of the type IBM PC, which function under control of operating system MS-DOS. The program uses standard functions of library Turbo Pascal: Ctr, DOS, Graph and also uncommon library of service programs of organizing inlet-outlet service, which are worked out by K.N. Zvagolskii and V.A. Gavrilenko.

As an example of program usage, we consider the concrete model: it is required to calculate the optimum heat exchanger in the system of heat providing of X-object, which warms initial water from 5 to 50°C. Cubic flow rate of the cold water is 5 m3/h. Limits of change of parameters for hot medium:

Temperature (input)	from 50,5 to 70°C
Heating flow rate	from 5 to 50 m3/h

The collection of inlet date is realized in structural file SUSEX DAT, where user,with the help of service screen editing, put the initial date for calculation. Screen form of file SUSEX DAT is presented in Fig. 4.10. The result of optimization is diagram of the "C-curve" in axes "adduced expenditures" – "SUSEX" and table of parameters of calculated points:

– Square of heat exchanger	– F	– sq. m
– Input (inlet) temperature of hot medium	– Tinp	– oC
– Expenditure of hot medium	– Vgr	– m3/h
– Number of heat exchanger	– No	– dimensionless
– Quantity of sections	– Nsec	– dimensionless
– Monetary expenses	– mon.ex	– rouble
– Sum of specific expenditures of exergy	– SUSEX	– dimensionless

OPTIMIZATION OF WATERHEATERS WITH C-CURVE WORKING

Version 1.0

Change limits of expenses Vgr =5 .. =50 m3/h
Change limits of temperature Tinp =50.5 .. =70°C

Vgr (var)
T1 (Tinp)°C ——→ T3 °C

Vx = 5 m3/h
T2 = 50 °C ——→ T4 =5 °C

Where T1, T2, Vgr-temperature of input, output and hot water consumption
T2, T4, Vc-temperature of input, output
and cold water consumption

Index of space Ispc=1
Electricity price = 30 roub/(MW•h)

1-hot water in interpipes space
2-cold water in interpipes space

Figure 4.10 Screen form of file of initial data

Table results of calculation, that was sorted according to criterion SUSEX, are put in the file of the final results SUSEX RES, which may be opened with the help of systematical commands MS-DOS.

Choice of the optimum variant is made by calculated points which are situated on the "arc of choice of the "C-curve". Rational quantity of a section of heat exchanger is taken into consideration (Fig. 4.11). In this example, the arc of choice is found by criterion SUSEX in the range from 1.49 to 1.53. This interval is found easily in the table of parameters of calculated points. The optimum variants of heat exchangers are chosen in it.

One of the variants we mean, for the heat exchanger N 8 (from standard line) with square 24.412 sq.m, has a quantity of section 7. According to the optimum expenditure of the hot water, if it is 7,4 m3/h, the optimum input temperature of the hot water is 57.7 oC (number of the point in the table of parameters of calculated points is 28), see table 4.1.

For comparison: if we hold optimization only from the point of view of minimization of monetary adduced expenses, the outcome of choice for heat exchanger will differ greatly from the previous one (input temperature of the hot water comes to Tmax).

Table 4.1

N	F m2	Tinp oC	Vgr m3/h	No –	Nsec –	Monetary rouble	SUSEX J/J
26	19.536	64.868	5.000	6	8.72	235.12974	1.49313
27	28.216	53.579	9.737	8	7.97	289.72023	1.49664
28	24.412	57.684	7.368	8	6.90	244.28072	1.50188
29	33.290	51.526	12.105	8	9.40	347.69321	1.51019
30	18.238	65.895	5.000	6	8.14	210.22750	1.52323
31	25.114	54.605	9.737	8	7.09	254.19116	1.52616
32	22.519	58.711	7.368	8	6.36	210.47769	1.53165

This program allows us to hold multivariant optimizations calculations of heat exchanging equipment uncommon by energetic and monetary criterion's in practical design.

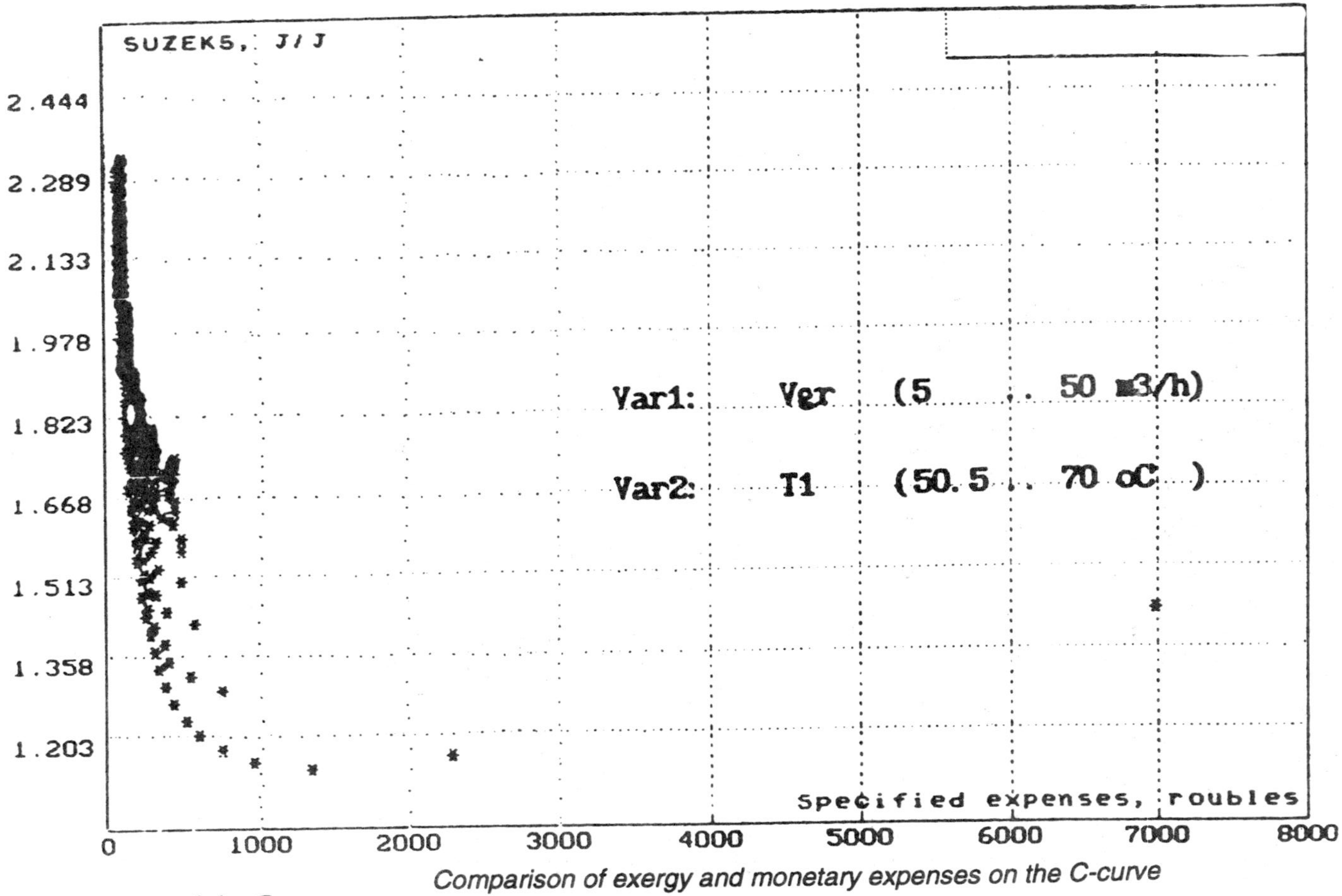

Figure 4.11 Diagram of the C-curve

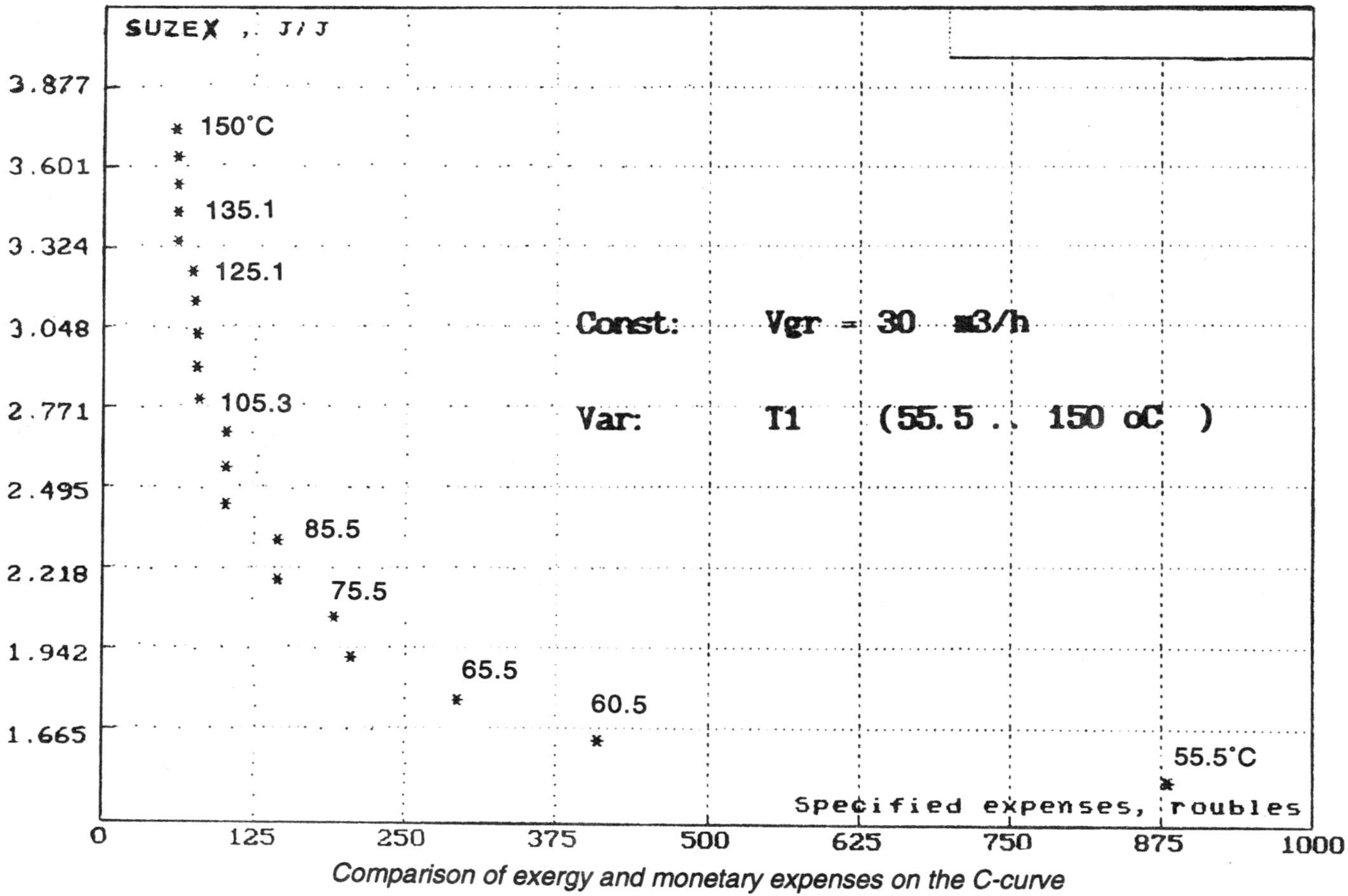

Figure 4.12 Diagram of the C-curve, variation of the heating water temperature.

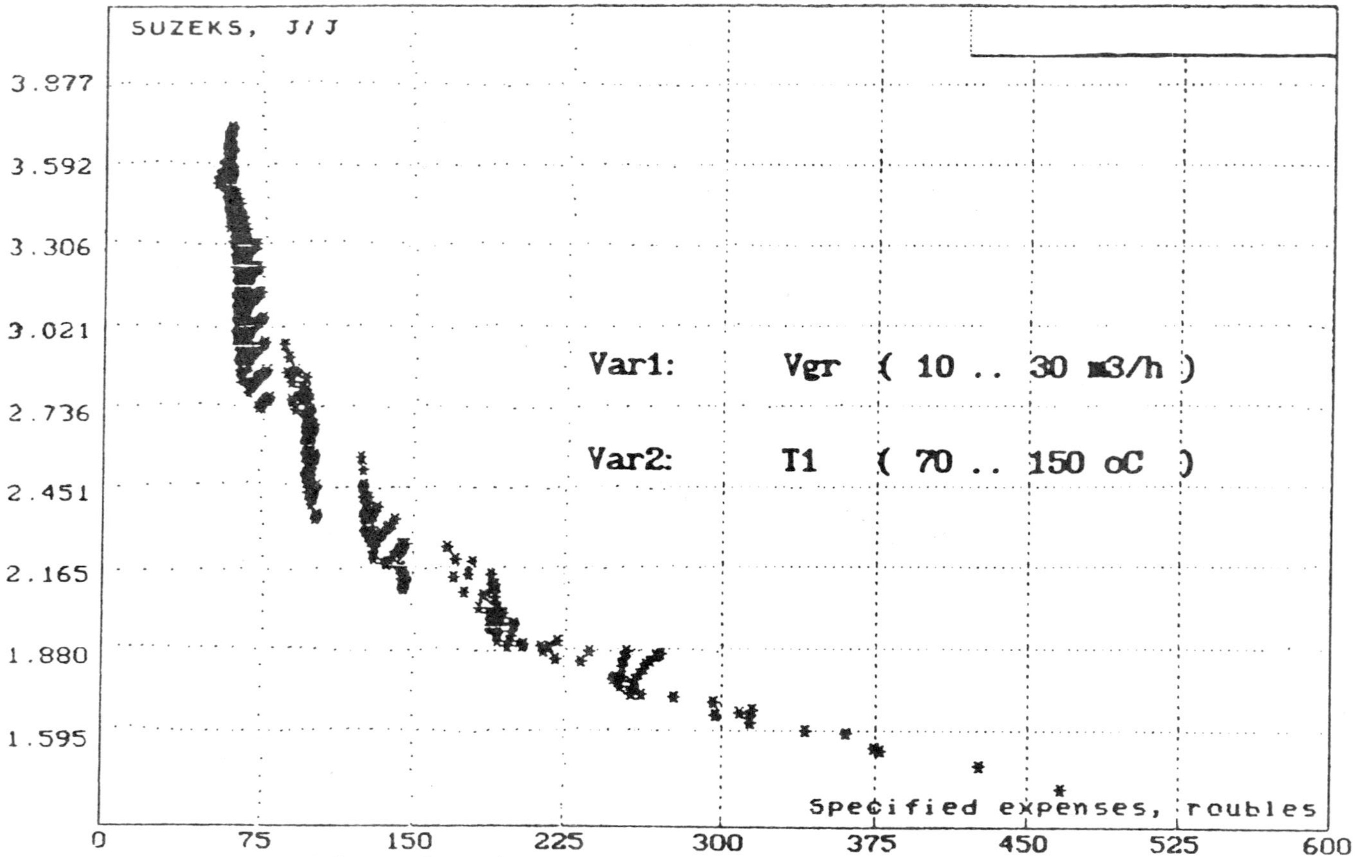

Comparison of exergy and monetary expenses on the C-curve

Figure 4.14 Diagram of the C-curve. Simultaneous variation of temperature and flow rate.

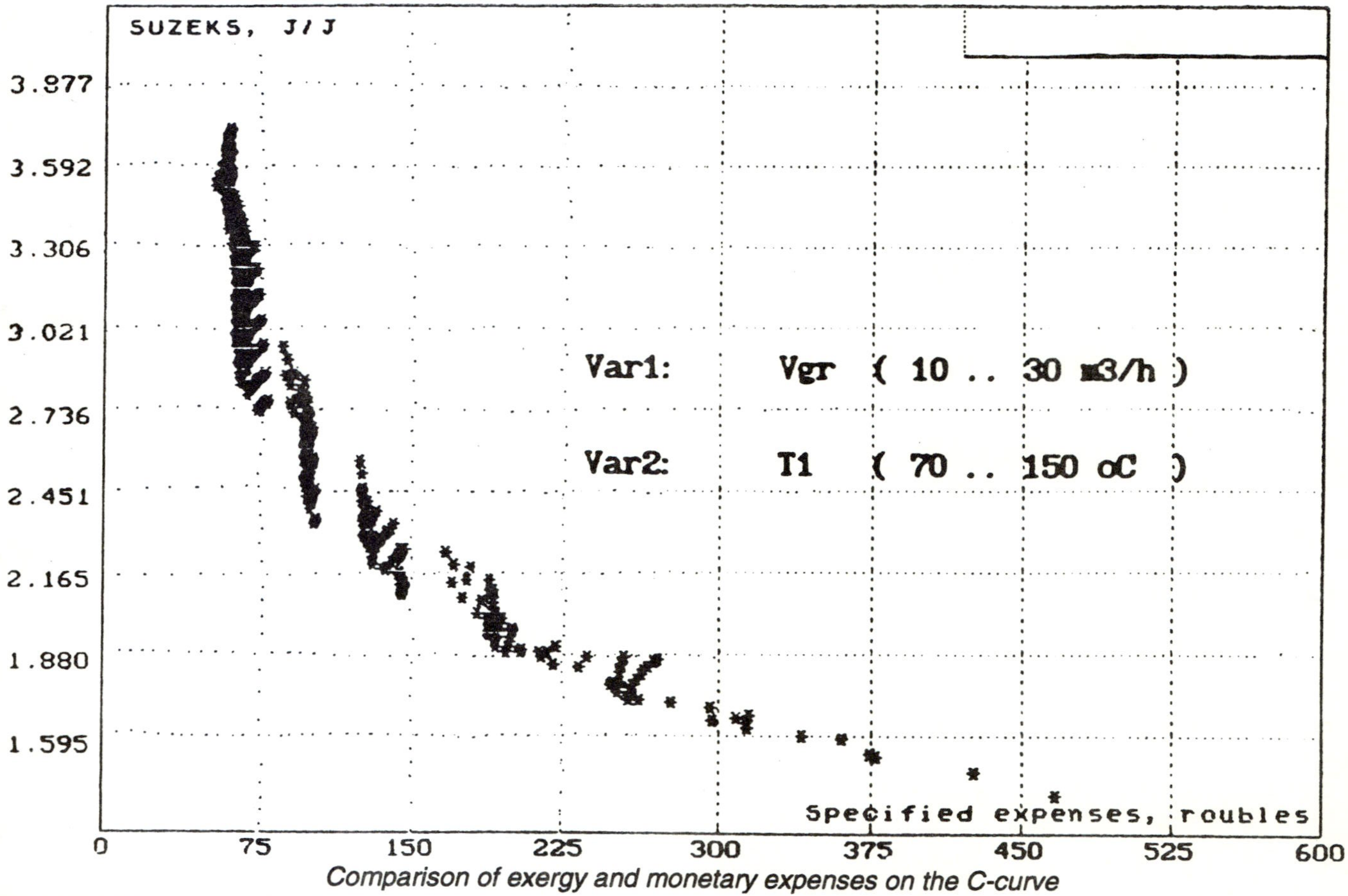

Figure 4.14 Diagram of the C-curve. Simultaneous variation of temperature and flow rate.

Chapter 5.

EXERGONOMICS OF SOME RENEWABLES

Do not forget that coal is the solar energy storage which was brought for millenniums. Consuming it in large quantities, we are putting in our budget random former capital while keeping ourselves as living on real income.

*If we were able to fix on the earth surface the same amount of solar energy, by that labor which was spent on the coal mining, the labor may be considered as useful indeed.**

S.A. Podolynskii, "Slovo", St. Petersburg, 1880.

Hitherto, we have taken the already stored solar wealth in fossil fuels and spent it ("living on capital"). Within 50 years, we shall have to be "catching" part of the incoming solar light and converting it to a useful energy medium ("living on income")...

It seems that a transition can only be attained by a tax on polluting fossil fuel to compensate for their pollutional damage, and redirection of this tax revenue in the building of the machinery for converting the renewable resources to usable clean energy.

J.O.M. Bockris, 7 World Hydrogen Energy Conf. Pergamon, v.1, 1988.

Payback Time

As a rule, the renewable energy sources are very material expensive. Because contemporary power production is based on fossil fuel, we can put to question if the renewable energy sources can conserve fossil fuels, taking into account energy investment. If the fuel expenditure needed to build a renewable is more than fuel saving, this project is senseless.

But the question may be put in more rigorous form. Because of energy discounting, payback time must be lesser than, not only the life time of the device, but lesser than the normative time span, nearly one fourth of life time (8 years).

---

* Translated from Russian by E.Y.

In this chapter, the total energy investment is calculated as the sum of energy consumptions on material production (ore excavation, melting, forging, etc.) which is the sum of products of the mass and specific energy consumption. The data for exergy intensity of materials has been collected by Szargut et al.

The energy consumption for the cold metal treatment and assembly of the details is minor and involved in the coefficient near unity.

The decision depends upon the inequality sign:

$$t_{pb} < t_{norm}$$
$$t_{pb} > t_{norm}$$

where t_{pb} could be obtained by equation:

$$\pi = \sum M_a \in_a + \frac{N}{\eta_a} t = \sum M_b \in_b \tag{5.1}$$

hence

$$t_{pb} = \eta_a \left(\sum m_b \in_b - \sum m_a \in_a \right)$$
$$m_a = M_a / N; \quad m_b = M_b / N \tag{5.2}$$

here N=electric power, the same in both cases for fossil fuel "a" and renewable sources "b", η_a=efficiency of the power plant, which is to be replaced, M=mass of different materials.

Neglecting energy investment on the replaced power "a" we have t_{pb} (in seconds)

$$t_{pb} = \eta_a \sum m_b \in_b \tag{5.3}$$

or

$$t_{pb} = \eta_a \sum m_b \in_b / 3600\ n \tag{5.4}$$

where t_{pb} is in years if n=working hours per year, $[m_b] = \frac{Kg}{W}; [\in] = \frac{J}{Kg}$

Let us consider an illustrative example; the OTEC (Ocean Thermal Energy Conversion) plant versus the gas fired power station (GFPS). Here n=7000 h/yr, η_a=0,4, N=100 MW.

The specific masses and energy intensities are as follows:

	OTEC		GFPS	
	m_b	ϵ_b	m_a	ϵ_a
Steel	1	60	0,2	60
Stainless Steel	0,6	80	0,05	80
Aluminum	0,4	300	0,05	300
Titanium	0,15	1300	0	0
Concrete	0		1	4

$$\text{Here } t_{pb} = \frac{0,4 \cdot 10^6}{3600 \cdot 7000} [(1 \cdot 60+0,6 \cdot 80+0,4 \cdot 300+0,15 \cdot 1300) -$$

$$-(0,2 \cdot 60+0,05 \cdot 80+0,05 \cdot 300+1 \cdot 4)] = \frac{0,4}{3,6 \cdot 7}(423-35) = 6,1 \text{ yr.}$$

If t_{norm}=8 years this t_{pb}=6,1 is marginally acceptable because uncertainty in involved figures. To firm decision t_{pb} is to be less.

When considering remote power installation on organic fuel transported with fuel consumption, it is necessary to diminish the η_a in proportion $1/(1+\gamma)$ where γ tonns of fuel is spent to deliver 1 t to user.

The described method was used to get a general outlook on renewables. Specific material consumptions data of Szargut, were recalculated in accordance to (5.3) and presented on Fig. 5.1. As it was in Winter (1989) paper the areas were numbered: 1=photovoltaic modules, 2=photovoltaic plants, 3=solar tower power plants, 4=wind power, 5=hydroelectric power. The solar pond power plant described in the next paragraph is depicted as black square 6. Its specific exergy consumption is the same as of a coal fired power plant.

Some new data on material intensity of the wind power (product guide) were used to calculate the payback periods of wind power in the range of 1 to 1000 kW; see black points on the Fig. 5.1. Previously estimated, OTEC is presented as the black triangle 8.

The physical reason of solar pond benefit is as follows: it is a wonderful concentrator of exergy flow. The falling on the pond surface averaged an exergy current density of 10^2 W/m^2, in contrast to the exergy current density in the hot brine from the pond bottom of 10^7 W/m^2 (five orders of magnitude more). The optical concentrators (mirrors, lenses, dishes) give a concentration of less than 10^3. In the stable working pond its hydrodynamic concentration is better than the optical in the mentioned devices.

The point 6 on the Fig. 5.1 relates to ordinary Rankine cycle steam turbine using the steam flow from the flash evaporator of the hot brine. Its payback time is by order of magnitude less than OTEC.

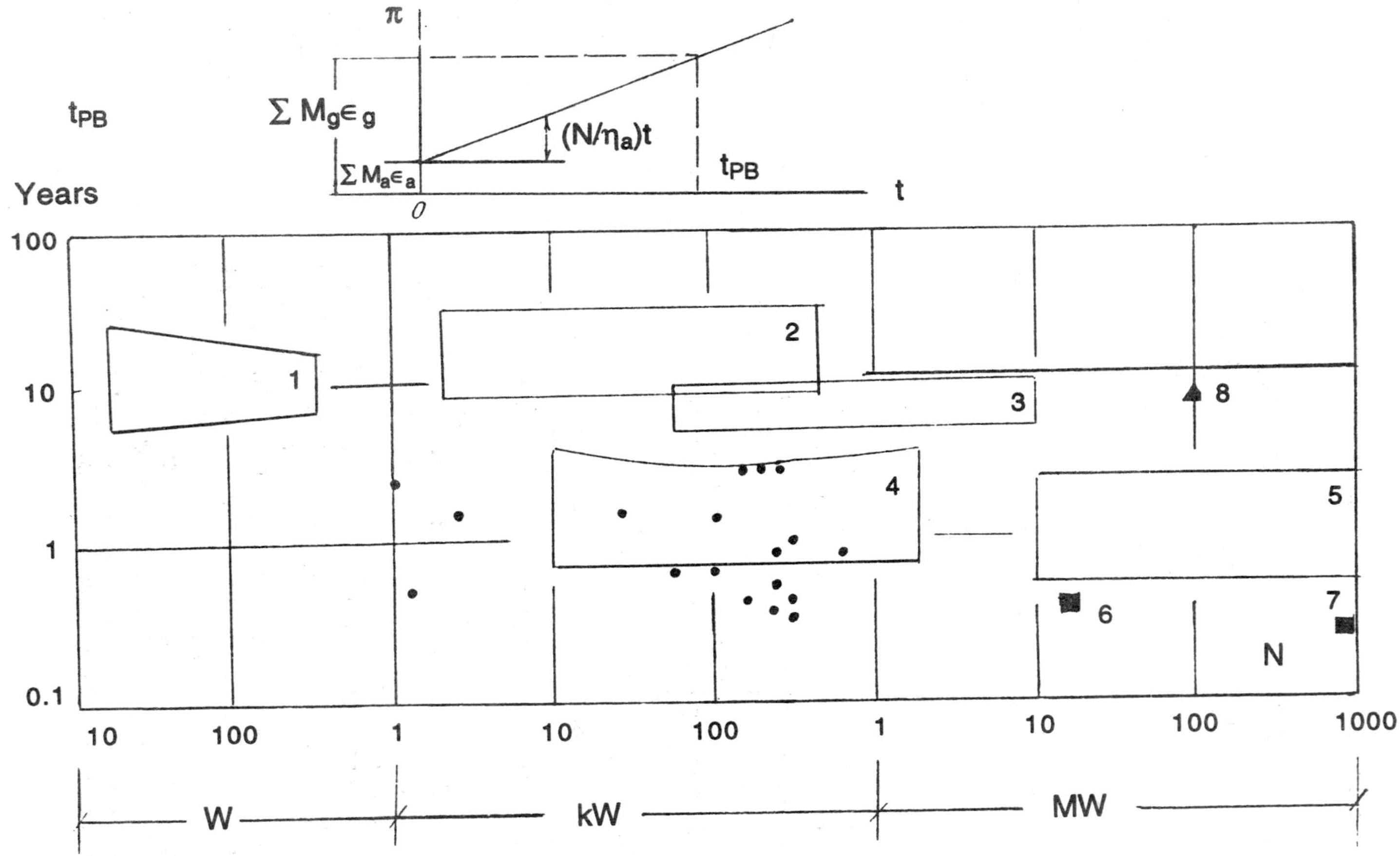

π
t_{PB}
$\sum M_g \epsilon_g$
$(N/\eta_a)t$
$\sum M_a \epsilon_a$
t_{PB}
t
0
Years
100
10
1
0.1
1
2
3
4
5
6
7
8
N
10
100
1
10
100
1
10
100
1000
W
kW
MW

Figure 5.1

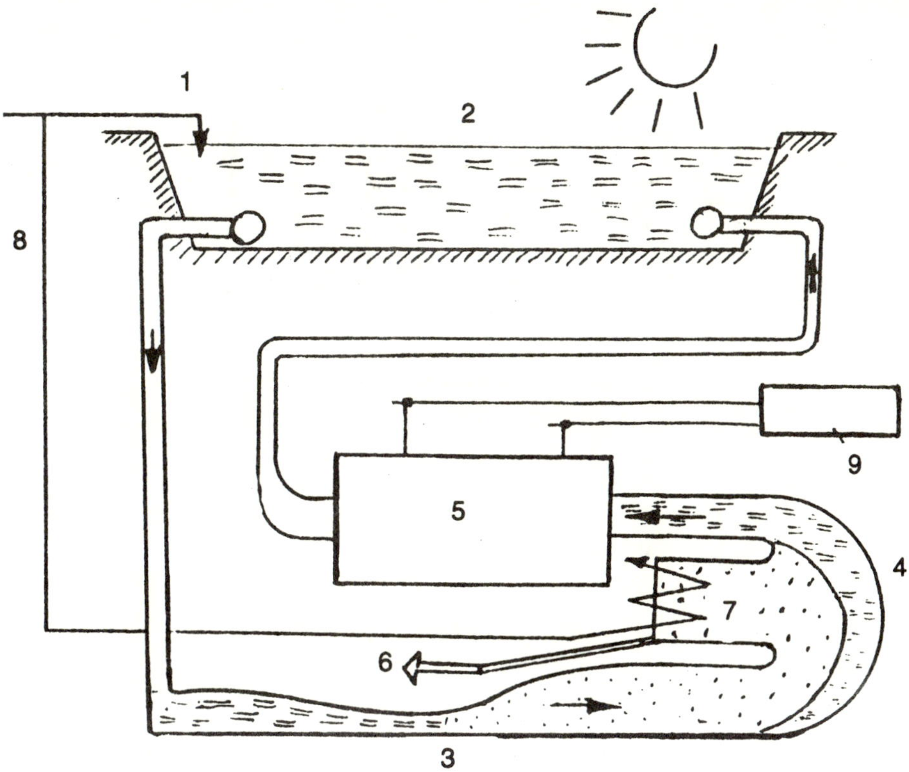

Figure 5.2 Schematic diagram of solar pond with MHD electrolyzer. 1-sea or make up water, 2-solar pond surface, 3-two-phase nozzle, 4-separator, 5-MHD generator, 6-condensed liquid, 7-condenser, 8-condensing water, 9-electrolyzer.

Here we may remember about high electrical conductivity of the strong brine (nearly 20 Sm/m) and the possibility to use MHD interaction for solar pond energy conversion. All the criterions of MHD interaction – the Hartmann number and Stewart number – depend upon dimensions of the MHD channel. Therefore, only large installations are possible instead of liquid metal MHD, where conductivity is 10^6 Sm/m.

The scheme MHD conversion of Lodhi is rather complicated because of intermediate working media – organics. The proposed scheme in Fig. 5.2 has no intermediate substances. There are no moving parts, even for brine return to the pond. Here we have: 1=make-up cold water (sea), 2=solar pond surface, 3=two-phase nozzle for brine self-acceleration, 4=inertial separator to set apart steam from the high velocity stream of conducting brine, 5=MHD generator, 6=fresh water consumers, 7=steam condenser, 8=cold water to condenser.

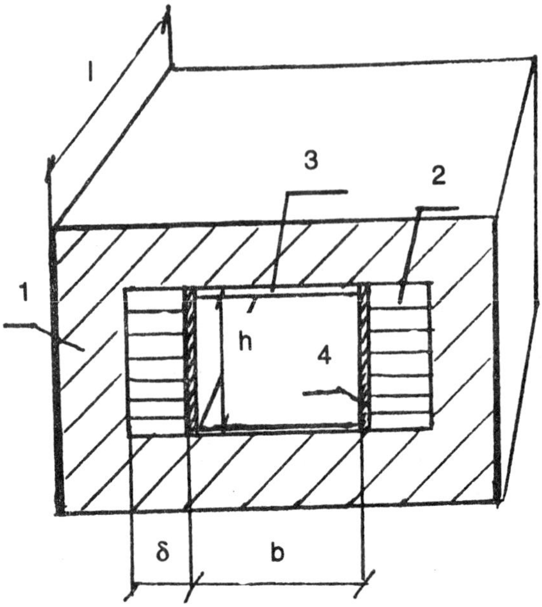

Figure 5.3 MHD generator. 1-iron core, 2-copper winding, 3-channel wall insulation, 4-short-circuited electrodes.

The cross-section of MHD generator is presented in the Fig. 5.3, where 1=steel core of the magnet system, 2=copper winding, 3=thin insulating wall, 4=electrode, b, h, 1=channel dimensions, δ=copper width. Electrical current through the brine in the channel leads to water splitting and hydrogen production. Here are different ways to use this internal MHD electrolysis on short-circuited electrodes, or to work on external electrical load – ordinary electrolyser.

The first crucial question for the SPMHD concerns the two-phase nozzle work. How much is the velocity of brine which could be achieved at the outlet of the separator?

The table 5.1 presents the results of the calculations (Kabakov et al) related to this problem. Initial pressures and temperatures of brine, which has to be accelerated changes in quite natural limits, 1÷2 bars and 363÷368K. The brine velocity after expansion varies from 134 to 147m/s, whereas that of steam varies 320-346 m/s.

The final result shows that for a large MHD generator, when flow rate of brine equals 100 ton/s the velocities after separator in the range 90÷97 m/s could be achieved.

For a rough approximation the figure 100 m/s is used in the evaluations of self-excitation of MHD generator.

The second crucial question in MHD conversion is the self-excitation; the loss of power in the electromagnet winding P must be less than generated power N, or P/N less one.

TABLE 5.1 INITIAL DATA AND CALCULATED RESULTS IN DETERMINING OUTLET VELOCITIES IN BOILING NOZZLE AND SEPARATOR ($P_2 = P_S = 0.005$ MPa, kinematio viscosity $0{,}76 \cdot 10^{-6}$ m/s) $G_1 = 100000$ kg/c

No	P_1 MPa	T_1 K	P_{1s} MPa	W_{2a} m/s	W'_{2a} m/s	W''_{2a} m/s	X_2	X_{2a}	ηn	Re x10^8	G'_1 x10^3kg/s	G''_1	C_f	W_s m/s
1	0.1	363	0,07011	134,02	106,95	320.86	0.0905	0.0714	0.45	66.92	92.86	7.14	0.001252	90.49
2	0.1	368	0.08545	147,84	115.23	345.69	0.0978	0.0807	0.47	72.09	91.93	8.07	0.001242	96.78
3	0.15	363	0.07041	134.2	107.1	321.31	0.0905	0.0714	0.45	67.01	92.86	7.14	0.001252	90.53
4	0.15	368	0.08451	148	115.35	346.06	0.0978	0.0807	0.47	72.17	91.93	8.07	0.001242	96.88
5	0.2	363	0.07011	134.02	107.09	321.28	0.0905	0.0714	0.45	67	92.86	7.14	0.001252	90.63
6	0.2	368	0.08451	145.13	115.46	346.37	0.0978	0.0807	0.47	72.24	91.39	8.07	0.001242	96.95

P_{1a} = initial brine pressure
T = initial brine temperature
P_{1s} = pressure after expansion
W_{2a} = actual flow velocity
W'_{2a} = liquidd phase
W''_{2a} = steam phase

X = Steam quality
η_n = nozzle efficiency
R_e = Reynolds number
G'_1 = Brine flow rate
G''_1 = Steam Flow rate
C_f = friction factor
W_s = velocity after Separator

Considering a simplified case of the constant cross-section of the channel and magnetic system we have (see Yantovskii, 1972):

$$N=k(k-1)\, \sigma_f V^2 B^2 bhl$$

$$P=I^2R=(j\, \delta\, h)^2\, 1/\sigma_W \delta h$$

$$B=\mu_0\, I/h$$

$$j=I/\delta\, h$$

$$P/N=[k(k-1)\ \mu_0^2 \sigma_f \sigma_W V^2 b\delta]^{-1}=\frac{1/k\ (k-1)}{(\mu_0 \sigma_f Vb)(\mu_o \sigma_W V\delta)}$$

Here we have; k=load parameter, and in denominator the product of the two magnetic Reynolds numbers - on the fluid conductivity σ_f and winding one. If k=0,8, b=4 m, δ=1 m, σ_f=20 Sm/m, $\sigma_W=4 \cdot 10^7$ Sm/m, V=100 m/sec we have:

$$\mu_0 \sigma_f Vb=1{,}25 \cdot 10^{-6} \cdot 20 \cdot 100 \cdot 4=1 \cdot 10^{-2},$$

$$\mu_0 \sigma_W V\delta=1{,}25 \cdot 10^{-6} \cdot 4 \cdot 10^7 \cdot 100 \cdot 1=5 \cdot 10^3,$$

$$P/N=1/0{,}16 \cdot 10^{-2} \cdot 5 \cdot 10^3=0{,}125<<1.$$

Hence, a large MHD generator on the brine stream may be self-exiting.

To minimize critical dimension we may remember about liquid nitrogen cooling of the winding, when the copper conductivity is ten times more. But the efficiency of nitrogen liquification is doubtful.

In principle, it is possible here to use the modern superconducting ceramics. Unfortunately the lack of information about the strength and energy intensity of these ceramics obstructs such net energy estimation. Net-energy is reliable for ordinary materials only (steel, copper, insulations).

In order to evaluate the net energy of such plant, the Faraday type MHD generator has been considered. Its channel inlet cross-section 1m^2, length 50 m, extracted power of 1000 MW.

The energy payback time was estimated as 0,2 yr, which is depicted as black square 7 on Fig. 5.1. It is less than of hydroelectric plants and OTEC.

The cost of hydrogen estimations (Bockris, 1988) show that wind, hydro and OTEC sources are better than solar thermodynamic and photovoltaics. Bearing in mind that exergy current density in the brine of solar pond is tenfold more than in OTEC, its low payback time appears reasonable and in agreement with Bockris.

In conclusion, one can see in the Fig. 5.1 that at first glance wind power and solar ponds have some benefit in comparison with other renewables.

See also Altfeld, et al.; Winter (1989); Schaefer; Lund et al.; Kreith et al.; Slesser.

Renewable Exergy Current Density World Champion

The main tool used in exergonomics is the exergy current density evaluation. As mentioned before, the equilibrium exergy is a rare appearance in practice because the currents are more frequently used instead. A list of currents was presented in Chapter 2 together with a useful division in convective and conductive parts. A definition of work and heat, as a conductive energy transfer by means of impulse and entropy currents, was given.

The major reason for using non-renewable energy resources is the very high concentration of chemical energy and exergy in the gas and oil pipes. The convective energy current density δ, in high pressure NG pipes, is of the order of 10^{11} W/m^2; the same is true for oil pipes. In HVDC electric lines, the energy current density (product of potential ϕ and electrical current density j_q) approaches the value of $750 \cdot 10^3 \cdot 1{,}3 \cdot 10^6 = 10^{12}$ W/m^2. Here we can see the main goal of a power plant: concentration in a power station from a boiler to electrical line reaches a factor of 10^8 (hundred millions).

Right here appears the major obstacle in using solar and other forms of renewable energy: the low energy and exergy current. The solar constant is 10^3W/m^2. The year-round figure for exergy current density (keeping in mind earth rotation, angles and clouds as well as entropy rise in radiation) does not exceed 10^2 W/m^2. The conductive energy transfer through the earth's lithosphere (geothermal heat) is even less. The convective energy transfer of geothermal energy in geysers is more concentrated, but the places where this happens are very few, and, under low (atmospheric) pressure, so the energy concentration is not high.

However, recently another source of geothermal energy was discovered, the "black smoker" (Podrajansky et al). The visitings on submersible apparatus to ocean floor in California Gulf gave a good chance to see very hot jets of strong brine in places of ocean floor that were 2km deep.

"Volcanism and hydrothermal activity keep the bottom sediments warm. According to data obtained from deep-sea drilling, the temperature in the abyssal layers of the sedimentary cover rises to as high as 300°C. The bathometer made out of titanium is ready to probe the water out of the 'black smoker' at the temperature of about 300°C. The bathometer preserves all that the hot solution contains, including the sediments that have precipitated in the process of cooling down."

At the depth of 200 m a jet of "black smoke" coming out of a piece of rock looks like a clay pitcher. The "black smoke" is a solution heated up to 350°C.

In the mouths of the powerful hydrothermal springs in depressions, the sediments form oddly-shaped debris, scoria, warped patches of crust and masses of fine-grained material. The cross-section of Guayamas through hydrothermal systems is presented in Fig. 5.4 Here 1=basement basalt; 2=hydrothermal sediments where primary solutions originate; 3=precipitants from solutions as they change on contact with sediments; 4=precipitants from solutions formed of thermic changes in sediments; 5=precipitant from solutions formed of changes in sediments upon contact with intruding molten basalts; 6=loose sediments; 7=sediments metamorphosed at 100°-300°C; 8=sediments metamorphosed by heat; and 9=hydrothermal simulation.

The most fascinating detail in Figure 5.4 for exergy specialists is the very close approach of isothermal lines (300°C to 100°C). These observations vividly show that in this particular case our Lord God in his infinite wisdom gives us the dream of exergy users: A cold and a hot heat reservoir next to each other. We trust also the scripture statement, "Will earth remaineth cold and heat shall not cease."

The following citations from Podrajansky et al. are interesting: "As to the fact that at the moment the camera came in contact with the 'black smoker' the ambient temperature was a mere 8°C, we put it down to the physical property of water. In one place its temperature was 300°C and a half a meter away it was down to 8°C. This means that the temperature of the "smoke" should be measured, not within half a meter of the vent, but inside the vent." And finally, "It is very dangerous to allow a jet of black smoke to lap over the sensitive equipment. In the case there would be nobody left to tell the sad story about what had actually happened to the viewpoint." Perhaps in the future this really heroic work of Soviet researchers would be highly appreciated.

But how about the exergy current density in the black smoke? Let us consider convective exergy transfer by this hot water jet with a simplified assumption. The strong brine specific heat is less than that of pure water, but we neglect this deviation because the density is higher; therefore, the volume specific heat capacity remains approximately the same. The velocity of jet could not be measured; it was estimated to be $2\frac{m}{s}$. Thus,

$$\delta_E \approx C\rho V\Delta t \cdot \frac{\Delta t}{T} = 4.18 \cdot 10^3 \cdot 10^3 \cdot 2 \cdot 300 \cdot \frac{300}{581} = 1.3 \cdot 10^9 \frac{W}{m^2}$$

Here Δt=300°C, T=8°C, V=$2\frac{m}{s}$.

The common controversy in exergy calculations about the reference temperature is meaningless, certainly it is the ocean deep water temperature of 8°C.

The above mentioned figure of δ_E=1,3 · 10^9 exceeds the solar constant by six orders of magnitude. This exergy current density is very close to the one in a gas pipe. Without doubt, this is the world champion of the natural currents of exergy density existing for millenniums.

Exergy Against Desert

In this paragraph, an exergonomic approach is used to illustrate the possible way of beating desertification, a harsh phenomenon in nature, today.

This is the struggle to improve living standard of densely populated regions of the planet, from equator to 45° of latitude, North and South.

The desert means the surface with rains less than 250 mm/year, where evaporation exceeds rain many times, the lack of surface water and impossibility of agriculture without irrigation. The deserts cover 83% of Australia, 59% of Africa, 38% of giant Asia and, all in all, 20% of land in the world. The most dangerous aspect is the annual increase of deserts by 50,000 sq km.

In some regions, human activity is the reason of desert increase. For example in Africa, the wooden fuel use reaches 3 cub. m per capita annually. The residential fuel use is 7-14 MJ/cap, which exceeds 5-7 times the highly developed countries. It is quite natural due to low efficiency of domestic stoves. The trees are already eliminated within 100km radius around African cities.

The deserts in Sudan moved, in 1976, south with velocity 5km/year. In Ethiopia deforestation reaches 1000 sq. km/y, which leads to droughts and desertification. Here the area suffering of desertification is guessed as 850 th. sq. km, in Sudan 2 mln sq. km, in Algeria 830 th. sq. km, in Mauritania 85% of land is desert, meanwhile Sahara is moving fast 6km/y.

A similar problem - Aral lake disaster, which is well documented in innumerous publications. The most clear and short description has been given by T. Kaipbergenov in 1987:

"In a quarter of century one generation of people has become eyewitness of drying of the sea, for merely of 60,000 sq km and 69 m in depth. Only a half remained. The Amu-Daria and Syr-Daria rivers may years enter nowhere... Seaport cities long ago are surrounded by sands only. Thousands of sq. km of green plants were dried out. Perspectives are very grim, ecological catastrophe is within sight. We have forgotten that the main function of a large body of water among the sands is to be natural solar absorber. As a result of sea drying, the salt which stored for centuries appears. Up to 75 mt of salt is gone with the wind annually. Nearly a billion tons of salt has been carried by the wind and transposed over the land. The coming catastrophe is not a defeat by the blind natural power, but the criminal behavior of people."

It is evident from all these data that the time has come for immediate action.

The tentative way to solve the above-mentioned problems is to use what resources which exist in the desert without shortages: insolation, land and salt. The aim of the approach is to get power and fresh water, just the things which are needed to stop the deserts' attack. The natural current of primary energy, solar one, in desert is quite high, region is cloudless. Maximum insolation reaches 10^3 W/sq.m, averaged 200-300 W/sq.m. The conversion of minor part of this energy, along with a fresh water gain, let us to stop the sands for sure. Here we remember the Oumov forecast for Sahara.

As we have seen already the mentioned figure 200-300 W/sq.m input energy current, even in the case of optical concentrators, lead to very low net exergy coefficient or large payback time. This obstacle, along with complexity of automatic control of mirrors, prevents the use of heliostats. The photovoltaics is more promising, but insufficient to very wide use.

To tap solar energy of deserts, we need the absorber of radiation which is created by nature, without energy intensive construction materials. The concentration of energy current by five orders of magnitude is needed in order to install the ordinary power plant equipment.

Such a solar energy convertor is a salt gradient solar pond. It stored solar energy in massive water hence this approach might be considered as heliohydropower, the match of solar engineering and hydro technique. The former gives information about solar intensity and variation, penetration of radiation into water, absorption and scattering of light with regards to wave length. The latter gives useful information on pumps, piping hot brine and water in ceramic and resin pipes, dam and pond construction.

Certainly the thermal engineering is needed here for new flash evaporators, turbines and compressors.

Salt Gradient Solar Pond

This is a natural or artificial body of water with sharp difference in salinity in the upper and lower layers. Salinity in the bottom reaches saturation, in good examples, when the temperature exceeds 100°C.

Physical background of such phenomena is the dumping of gravitational (free) convection, in other words, prevention of rising heated water up to the surface of the pond, where heat might be transferred to air, especially in windy time. Heavy strong brine in the bottom zone cannot rise because, in spite of lessening the weight due to heating and thermal expansion, the increase of specific weight takes place as a result of the dissolving of heavy salt. The increase of weight due to salt exceeds decrease due to thermal expansion.

The infrared part of solar beam used to be absorbed in the thin upper layer at one centimeter deep and without any doubts is gone with the wind. However, the major part of solar energy is transferred by ultraviolet and short visual band waves, which might penetrate into transparent water to some meters depth. It is absorbed by the floor of pond then the two flows of heat conduction takes place: up to surface through the steady-state gradient layer of water and down, through the thermally conducting ground to underground water table, if any.

In spite of this heat loss, the temperature 102°C and 109°C were registered and some theoretical guesses of 150°C in saturated solutions as well.

Certainly it depends upon the latitude. A pond consists of an upper (convective) zone 0,1÷0,3 m then, deeper, nonconvective stable layer with halogradient and thermogradient. The depth of this layer influences on all the properties of the pond, especially the deepest-bottom convective zone, where energy is stored. The thermal inertia of the pond heavily depends upon bottom zone thickness.

The pond efficiency is daily averaged by the ratio of extracted heat per unit surface and falling solar energy current.

The great Russian chemist D.I. Mendeleev wrote in 1889, in the book *Background of Chemistry* many data on salt pond technology and construction: "Sodium chloride, or well known to any layman as salt, exists in the earth core, dissolved and extracted by atmospheric water, flowing to seas and oceans.

In the Earth's long history, this process has accumulated in oceans and a lot of salts were formed due to water evaporation.

Sea water salt serves as the source for its commercial extraction, as well as solid salt, salt wells and lakes.

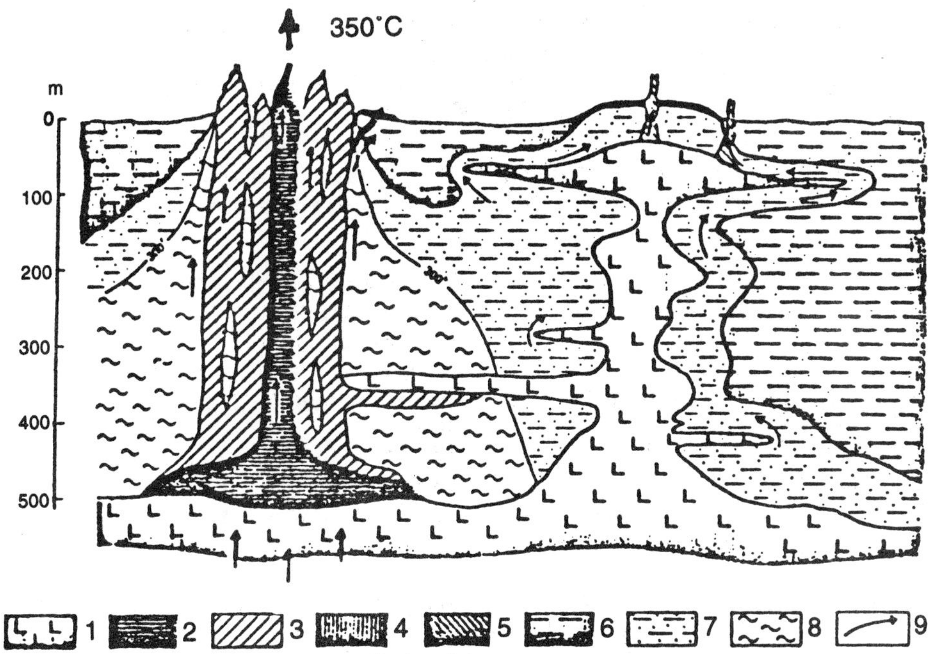

Figure 5.4 Cross section of a hydrothermal system.

There exist several ways to extract salt of sea water. In southern countries, on the coastal regions of Atlantics, Mediterranean and Black Seas, the summer hot weather is used. Low and flat seashores are selected and the series of artificial ponds are built, connected in between. Sea water is pumped or directed when it is high tide to upper basin. The artificial or natural dams is what separates these ponds from the seas. In April, significant evaporation begins. The brine is conducted to the next pond continuously or stepwise, and so over the total series. The bed of the pond has to be lined to prevent water leakage. For this purpose the clay bed is to be firmly pressed. When salt content reaches 28%, the crystals of salt appear, which are bulldozed and then consumed. In most cases only the first half of salt deposit is used because the second part is bitter due to magnesium salts."

After these long citations, we have to note that just these wastes or residual brines of salt plants are, in modern times, the most commonly used working media for testing of solar ponds.

The thermal stability of a pond is profound, its temperature dependence upon clouds and hours in a day is quite weak.

Even at the latitude of Middle Asia (40°C), in the winter, temperature doesn't fall below 50°C as calculations showed.

Hot brine from the pond may be set apart on the fresh water (distillate) and strong brine in the combined power – desalination plant, consisting of flash chamber (evaporator), steam low pressure turbine and condenser (Fig. 5.5):

1 - Solar pond
2 - Ceramic out flow pipe
3 - Ceramic in flow pipe
4 - Brine pump
5 - Flash evaporator
6 - Steam turbine
7 - AC generator to grid
8 - DC generator for electrolize
9 - Electrolyzer for hydrogen production
10 - Hydrogen high pressure store
11 - Shell and tube condenser
12 - Fresh water for consumers
13 - Heat accumulator
14 - District heating source
15 - Cold water source (sea)
16 - Sea water pump
17 - High pressure steam

The hot brine from the bottom of the pond 1 is exhausted by pipe 2 to evaporator 5, where low pressure (below 1 ata) is maintained. Clean steam by 0,7-0,9 ata through turbine 6, is expanded to 0,03 ata and condensed in 11 by cold water from 15. Turbine power is transferred to grid by alternator 7. Direct current from generator 8 is used to split brine on hydrogen and oxygen in electrolyzer 9. Hydrogen is accumulated in tanks 10 or other, for example, hydride accumulators. Fresh water 12 is pumped to consumers. Part of the heat is directed to heat accumulator 13 for low grade heat consumers. High potential heat (high pressure steam) is delivered by compressor 17, realizing the open cycle heat pump.

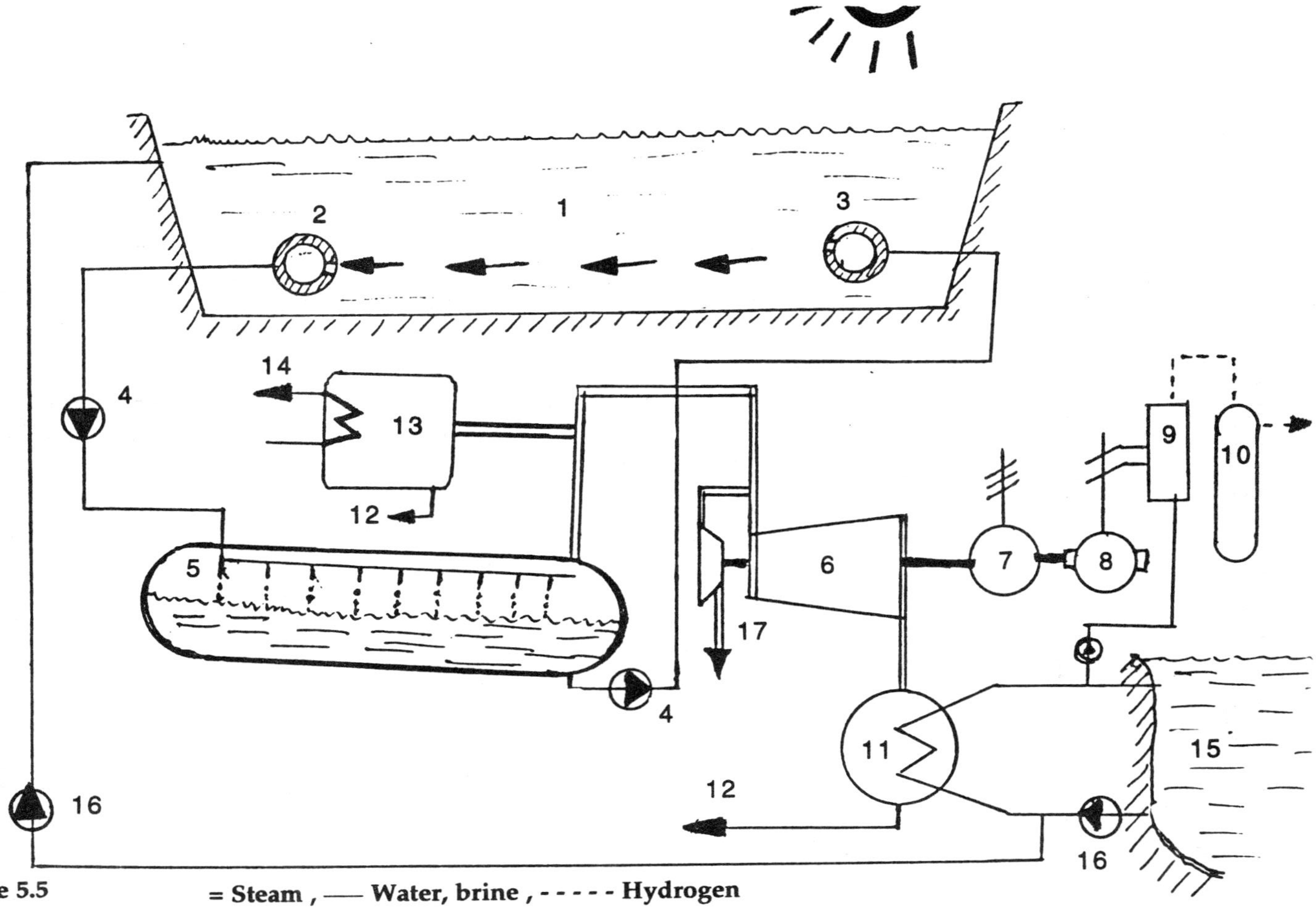

Figure 5.5 = **Steam** , —— **Water, brine** , - - - - - **Hydrogen**

An MHD generator on the brine flow with electrical conductivity by order of magnitude greater than sea water wasn't shown on the picture 5.5 (see Figures 5.2 and 5.3).

Without an external load short circuited an MHD generator is turned to electrolyzer having hydrogen yield on one electrode. The brine flows by means of two-phase nozzle, well known from the experience of liquid-metal MHD generators.

In the middle of summer, maximum possible parameters of a pond with sodium chloride or magnesium chloride are as follows:

Insolation - 250 w/m^2
Brine temperature - 100°C
Cold water temperature - 8°C

Because of the loss of heat from the bottom zone of the pond through thermal conductivity (to the atmosphere from above and to the ground from below) the efficiency of the pond as solar energy converter is 0,15-0,30.

For maximum boiling temperature should reach 94°C and condensation temperature is 16°C, the Carnot efficiency is (94-16)/(94+273)=0,214. Let exergetic efficiency be 0,5, then maximum total efficiency is to be 0,1 and maximum converted energy flux density is 250 0,1·0,3=7,5 W/m^2.

Lowering pond efficiency, insolation and brine quality diminishes this figure to 4-5 W/m^2. For the summer-averaged condition at the 40-45°N latitude there may be fixed power density 4W/m^2 and for year averaged 2-2,5 W/m^2.

In other words, anticipated full efficiency of solar-electric energy conversion is about 2%. It is by order of magnitude less than in the best photovoltaic converter. However energy expenditure on the solar pond is many orders of magnitude less than photovoltaics giving the way to ponds.

The net-energy calculations evaluating effectiveness of the solar pond power (without hydrogen part) were done. The low pressure part of the ordinary turbine BK-100 with steam flow nearly 300 t/h was taken.

For different temperature drop in brine flow through flash chamber (2-13°C) the narrow optimum at 6° was obtained. The power output and the sum of specific exergy expenditure versus above mentioned temperature drop are depicted on the Fig. 5.6 for magnesium chloride (34%) and sodium chloride (20%). The distance from the pond - 1km and from the cold water source - 5km. These figures are important because of the steel for water pipe. Nominal output BK-100 LP part is near 30 MW. In our case, gain from the pond is 15-20 MW over the pump loss of power 5 MW.

The site of installation was found in the extreme southern part of the Sivash salt lake in Crimea, taking into account the very near location of cold water in the deep of Pheodossia Bay.

If the half of the surface of Sivash is used, it will give nearly 5 GW (e) and 60 m^3/s fresh water. It will be a radical solution to energy and water problems of the Crimea and the Azov sea.

Specific exergy expenditure (SUSEX) about 3 is the same as for coal-fired power plant. If to compare solar averaged exergy flux density (200

W/m²) and the same flux density in the ceramic pipe with hot brine ($c\rho V\Delta T^2/T$) one can see concentration of exergy by factor 10^5. This is a crucial benefit of the solar pond concept.

Experimental researches were carried out on the small ponds some tens of years ago, beginning from the first observations of Russian scientist A.V. Kalechinsky in 1901 on the Madve Lake in Transilvania. Interesting results were obtained by Yu. Usmanov in Tashkent (90°C in 1970).

The most interesting plant (project power 5000 kWe, closed cycle on freon) was tested in July 1984 (H. Tabor) on the Dead Sea region. The economics of the designed plant is very attractive – the cost of energy, 20 cents/kwh by achieved 1,25 Mw and 9,5 cents/kWh by 10 Mw. The product of pond efficiency 0,2 and cycle efficiency 0,07 equals 1,4%. The achieved full efficiency of solar-electric conversion is about 1%. In the recent review of Abdel Salam many pond experiments and constructions were described, (see also Posnansky.)

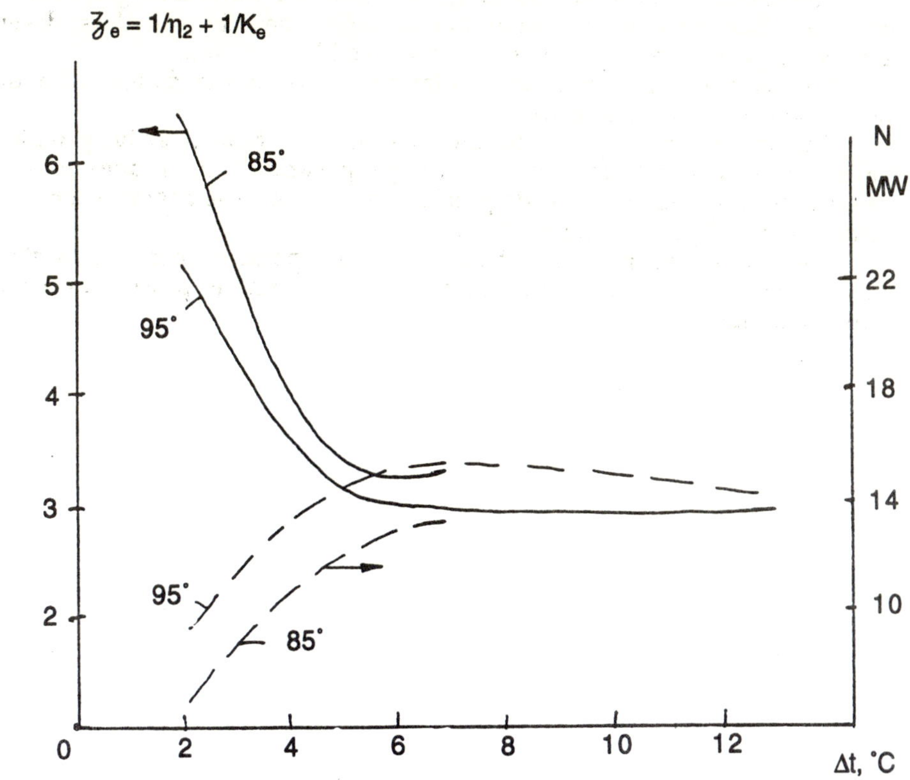

Figure 5.6

As it was well known, solar hydrogen energy is closed by the flow of substance – water at the start and the same water at the end as a product of real or "cold" combustion in fuel cells. This is the main benefit of SHPE (Solar-Hydrogen Pond Energy) in contrast with nuclear power with unavoidable radioactive wastes and the wastes from fossil fuel plants.

In contrast with all the fuel power schemes, SHPE does not emit any substance which was absent in the nature before.

In comparison with hydraulic turbine power, SHPE is preferable because of less water storage needed (by volume of brine store). As is well known heating up the water (and brine too) on 1K equals the height of 400 M. The exergy by 94°C equals 0,2 of enthalpy, so 1K means 80 m height. When one cool brine on 10°C gains 800 m, that is three times higher than Nurek dam and 40 times higher in comparison with dams on flat plate rivers.

The surface of a pond may be greater than the water storage (when not so deep). But the cost of land is not so high when the salt flats, deserts, artificial salt lakes and so on, are without arable land loss.

The lack of high dams on a solar pond avoids the danger of a dam failure and other catastrophes.

The salt technology is the oldest branch of human activity with no harm for its surroundings. It is benign to people and nature. It does not cause smogs, acid rains, carbon dioxide problems and so on. The only problem is the nitric oxides in hydrogen combustion.

We will see soon, that salt gradient solar pond concept has a strong competitor, the SOFT concept, based on photosynthesis process in a pond with microalgae.

Chapter 6.

CLEAN POWER CYCLES

They don't throw their garbage away. They make it into TV shows.
W. Allen, M. Brickman

Short Review

In ancient times, the contents of night pots in some European countries was thrown out of windows into the streets. After this practice was forbidden, some citizens used to stealthily at night do the same over the fences of their neighbors. The very high chimneys of our fuel-fired power plants are in effect devices for disposing of waste gases and particulate matter to our neighbors who tolerate this and behave likewise.

However, recently it was recognized that as a result there is a greenhouse effect and all the neighbors have become the inhabitants of one apartment. The greenhouse effect is a global problem indeed; and unfortunately, great investments are needed now in order to protect the future. Till now little progress has been made, in spite of strict environmental regulations.

The time has come to understand that fossil-fuel power process should not end in the atmosphere. We urgently need to prolong the ways of effluents to bypass the atmosphere and to finish the trajectory in the depth of open or underground ocean.

Looking at power plants, a real start might be made if we recognize not only what the problem is, but also what the solution should be. We must find for future power plants not only the cycle which is economically attractive, providing substantial energy conservation; but one which is also ecologically benign.

The extraordinary scale of economic impact of the global warming due to greenhouse effect has been demonstrated by Laurmann (1991) according to A.S. Mann and R.G. Richels: emission reduction costs for USA are guessed in four trillions of dollars.

The review of Blok et al. (1991) shows a large amount of different approaches to the CO_2 removal problem. By our point of view, the most promising is the modification of the energy conversion process including combustion in the oxygen/flue gas mixture. The firing of coal in such a mixture has been investigated in the ANL USA by Abele et al, Wolsky et al., see Blok review. The decline of efficiency from 35% to 24-26% in the case of zero CO_2 emission was registered. Coal gasification in the IGCC, on the base of design made in Fluor Daniel with the Shell gasifier and GE gas turbines, has been considered in detail by Hendriks et al. (1991). The efficiency decline from 43.6% to 38.1% was calculated, the CO_2 emission was reduced from 0.75 to 0.13 kg/kWh.

H. Hertzog, D. Golomb and S. Zemba of MIT recently considered many CO_2 emission-free power plants with sequestering all the effluents in the deep ocean. They mentioned the first proposal of C. Marchetti (1977) on disposing CO_2 in ocean by means of big 1000 km pipeline.

Among others, they mentioned also the Hendriks article on the Integrated Gasification Combined Cycle (IGCC) power plant with physical absorption by selexol. This plant's thermal efficiency falls from 43.6% to 38,1% due to CO_2 capture. Hendriks et al. proposed to sequester the CO_2 from Dutch power plants in the exhausted North Sea gas wells.

Hertzog et al. have brought together the results of analysis in main US laboratories, giving the share of coal energy, consumed to get oxygen and compress CO_2:

	Air separating	Compression
MIT	17	9-14
BNL	16	9
ANL	45	--
	--	16
ORNL	32	5

The figure for oxygen production energy consumption, as 200 kWh per ton is recommended in view of the future and for very big air separation plants of 100 kg O_2/sec.

These authors have restricted themselves within ordinary power plants with boilers, replacing in some cases the nitrogen by CO_2 as a heat carrier. The principal aim of the present chapter is to consider some new cycles, where efficiency does not decline and even might increase in spite of zero atmospheric emissions. In the ANL scheme, combustion was external with respect to the cycle working substance. In our case the combustion is internal, without energy transfer through the wall of boiler.

The logic of investigation is based on exergy flow concept, sometimes hidden, sometimes explicit. We should understand that cleanliness does not mean the lack of exergy losses. For example, the more is turbine inlet temperature in a gas turbine unit the less is exergy loss, however the more is toxic NOx release. In the internal combustion cycles, we are dealing with, there are no restrictions to exergy efficiency rise.

D.P. Hochstein proposed in 1940 the external combustion CO_2 cycle as more efficient than the steam one. The idea gradually was going ahead, up to the project of 50 MW demonstration plant, see Hochstein and Kirov (1983). Unfortunately the work has been cancelled due to the death of its originator.

P. Pechtl has considered some schemes with internal combustion gas turbine and CO_2 as the working substance of the cycle. The coal-fired IGCC power plant efficiency is evaluated as 38,9% by 500 MW.

The CO_2 here is also disposed into the ocean depths, 760 m. The liquefaction of 414,8 ton/hour of CO_2 needs 26,5 MW power (5,3% of generated). Efficiency falls to 36%.

P. Pechtl estimated the demand of CO_2 for the chemical industry as 1,6% of its content in stack gases of power plants by mass of carbon. Therefore, it can not be reused in industry and must be carefully disposed.

Pechtl considers also the empty oil wells as CO_2 storehouses and the returned mass is 2,6 times more than oil with the same carbon content.

Some new hopes appear, if not restrict, the analysis by existing plants, and considering the near future to find another site for effluents disposition, and other ways not to lose but to gain in efficiency.

W. Seifritz has proposed a new energy system whereby primary fossil fuel, like coal or natural gas can be utilized together with a modest use of nuclear high temperature reactor without any release of CO_2 (and other chemical contaminants) into the atmosphere, avoiding greenhouse effects on this planet. Natural gas is converted into methanol which is an energy carrier. After decomposition, the energy hidden in the C-atom is converted in CO-turbines into electricity, whereas the energy hidden in the H-atoms is used in decentralized manner. Many previous papers about carbon dioxide removal, recovery and disposal, especially of M. Steinberg, were quoted.

Seifritz vividly shows the impossibility of CO_2 disposal into space, because shooting of 1 g needs 63 kJ, which greatly exceeds the formation energy 9,25 kJ. Therefore, CO_2 has to be disposed within the planet.

Seifritz has considered only one way to CO_2 effluents, the ocean deeps, including ingenious invention of CO_2-Ice "Torpedos". The flow sheet of CO-fired in oxygen turbine plant, with air splitting unit at the entrance and liquid CO_2 pipeline at the exit, is presented. Some data about ocean as a CO_2 storehouse will be given later, on p. 89.

Underground Waters as CO_2 Absorber

There are many small countries located far from the sea coast, as well as big ones, where power plants can not be near to a shore. In some cases there exists underground water or brine.

For example, nearly 500 cub. km of brines exist in the small area in between rivers Volga and Ural in the center of Russia at the depth of 20-25 m. On the full South of the USSR, many tens of thousands of cub. km of brines are registered (Zaitsev). Nearly the same waters exist in other countries.

In view of these waters, two natural questions are as follows:

- How much of CO_2 might be dissolved, and,
- What is the fate of this solution?

The first answer we can get by means of Fig. 6.1, where some data on the solubility of CO_2 in sea water are presented. The equilibrium content in normal pressure (0,1%) is quite low. This lets us separate gaseous CO_2 from water in the considered cycle. However, by high pressures, as in ocean depths, more than 38 bars, the CO_2 share exceeds 0,06 (6% by weight). The very high pressure exists in underground brines, therefore, in our region example in between rivers Volga and Ural, there might be stored more than 30 Gigatons of CO_2, which greatly exceeds annual world CO_2 emission.

In view of the mentioned tens of thousands of cub. km, the capacity of underground brines to meet man-made CO_2 is evident. In some places for this purpose drilled dry oil wells might be used.

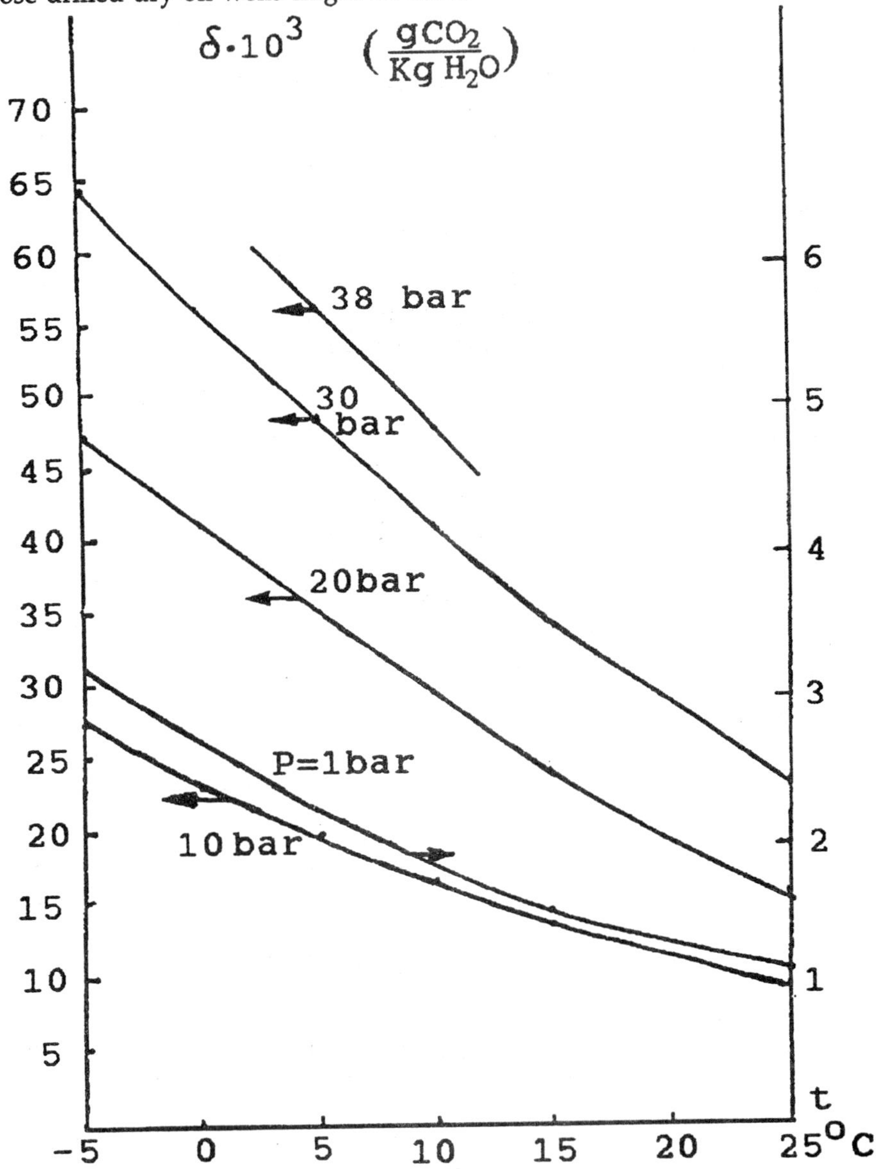

Figure 6.1 Equilibrium contents of CO_2, dissolved in sea water (Stewart P.B.).

There exists a growing consensus that hydrolithosphera is the best storehouse for CO_2 disposal. Deep underground aquifiers are quite widespread. The depleted oil and gas wells seem to be the less expensive way to reach it (van Engelenburg and Blok, Holt and Lindeberg).

Let us estimate the ultimate absorbing capacity of underground brines. According to Makarenko, mentioned by Perelman (1983), the global water balance consists of:

Oceans 13700 gg, Underground 3300 gg, Glaciers 200 gg.

Here gg=geogramm=100 th. gigaton. If we assume the solubility of CO_2 in water as 10% and global man-made CO_2 release as 33 Gt/y, the time span is million of years.

As to the fate of CO_2 in solutions, the carbon acid formation should be kept in mind, therefore the reaction of alkaline neutralization is expected.

The $CaCO_3$ formation of water, CO_2 and fluorite and the $MgCO_3$ of water and CO_2 by very high pressure and temperature in the planet crust, is considered in (Nordstrem, pp. 246–257). In any case this carbon dioxide never enters the atmosphere.

In some particular cases, even on the surface, the strong alkaline lakes and ponds have appeared, for instance, ash ponds of the shale-fired power plants. It is the good place to dissolve CO_2, to neutralize ponds getting carbonates and, last but not least, to return life in these basins.

The problem of CO_2 removal and sequestering has been discussed in details at Amsterdam Conference, 1992, March 4-6 (see Blok, 1992).

Steam Internal Combustion Power Cycles

In principle, only two inert gases might be used for the replacing of nitrogen before combustion, namely steam and carbon dioxide. They are used as working substances of the considered cycles, see Fig. 6.2. The external combustion cycles need no explanation. These are ordinary Rankins or Brayton cycles with external heat sources.

The schemes which are considered are

- the GOOSTWEG power cycle; gas or oil fired oxygen steam turbine without exhaust gases – recommended for gaseous or liquid fuels
- COMHDWEG power cycle; coal fired oxygen magnetohydrodynamic generator without exhaust gases – recommended for coal or low grade fuels such as shale oil. Here the combusion product is heavily CO_2.
- COGSTWEG power cycle; coal-oxygen gasified steam turbine without exhaust gases – recommended for coal and shale oil use
- SOFT power cycle; solar oxygen fuel turbine for solar energy use.

All these cycles serve one primary idea which is to eliminate the release of harmful exhaust gases by means of the elimination of nitrogen before combustion. Certainly the simplest case is with the use of natural gas. The author believes that there is at least more than a hundred years supply of this fuel left in the world in view of the off-shore and hydrate resources which are available. Many countries such as Russia, the United States, the United Kingdom, the Netherlands and others will use it for a long time.

The coal option is also important. However, in this case pollution, which is the main problem, must be strictly prevented.

The last type of fuel, solar energy, as we have seen already, is the ultimate fuel for the development of power production. Here the SOFT cycle seems to be extremely useful. According to Professor John Bockris (1990):

"The major difficulty is that the efficiency of light conversion to energy via biomass is small (about 1%), and most countries do not have sufficient free land-area on which to accommodate the necessary trees or other crops. If one could use genetic manipulation to make biomass grow at a 10% per year energy-conversion efficiency, then the possibility of biomass providing a solution would be greatly advanced."

The SOFT cycle involves well developed photosynthesis with just 10% efficiency by means of ponds with microalgae.

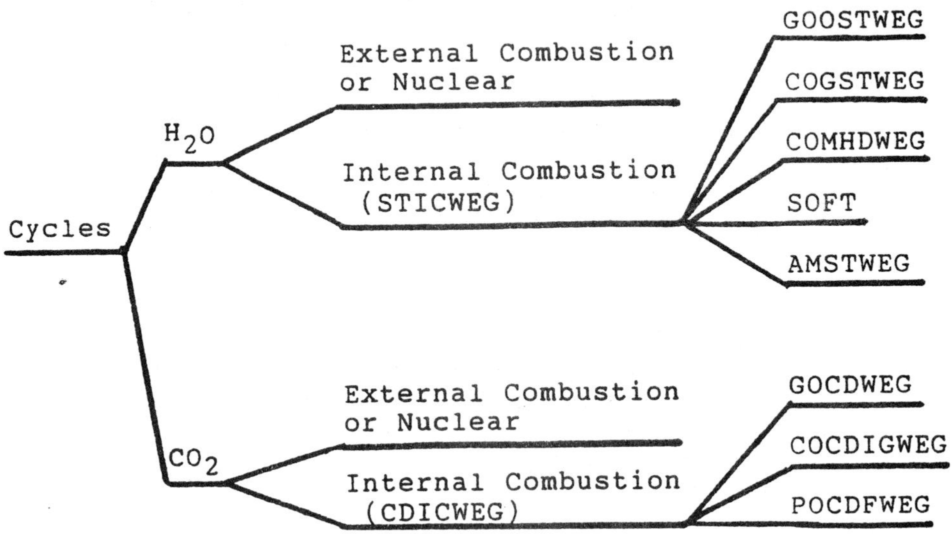

Figure 6.2 Thermodynamic cycles hierarchy

The GOOSTWEG Cycle

The plant outline is presented in Figure 6.3. Air enters splitting machine 1 where it is divided into nitrogen and oxygen. The former is released to the atmosphere without any harm or is used elsewhere. Oxygen mixed with high pressure feedwater in 2 enters combustion chamber 3 to which compressed methane is directed. Steam injection into the combustion chambers of the gas turbines is widely used, for example, by General Electric in its STIG cycles. Our case differs due to the almost complete elimination of nitrogen before combustion.

In the excellent book of J. Bockris (1980, p. 264) one can find the scheme of Aphodid burner steam plant, where hydrogen is combusted in mixture of pure oxygen with excess cooling water (US patent of Hughes and Brauser, 1969). Here combustion product is just pure steam driving ordinary turbine. Hydrogen is still an exotic fuel. In our case the most abundant fossile fuel, natural or coal-derived gas is used.

The temperature prior to the turbines is 750°C and is enough to get acceptable efficiencies without special blade cooling. Possible increases in temperature up to those used in modern gas turbines (up to 1300°C) are a chance for future improvements. This will be shown later.

After triple heating and expansion in turbines 4, 6 and 8, the mixture of steam and CO_2 enters recuperator 9, heating the feedwater, fuel and oxygen in separate heat exchangers. The steam is then condensed in the condenser and returned. The carbon dioxide from the degasser enters the compressor and in liquid form is disposed.

In the calculations, the well known M.P. Wukalovich Tables of the thermodynamic properties of water and steam were used (seventh edition, Moscow, Energy, 1965), with some corrections for the addition of CO_2. Note, however, that there is some degree of uncertainty in these corrections.

We admit this inaccuracy in the preliminary guesses only, to show the logic of calculations. The most promising cycles as SOFT and COOPERATE in the next paragraphs have been computer simulated. There the properties of real mixtures of flue gases are calculated simultaneously.

The cycle is presented in Fig. 6.4 in t-S coordinates. Its principle state points are as follows:

Table 6.1 Thermodynamic state points for the GOOSTWEG cycle

Point	T (°C)	P(bars)	s(kJ/kg K)	i(kJ/kg)
a	80.	.5	1.1	.33
b		240.	1.1	.39
c	200.	240.	2.3	.86
d	750.	240.	6.8	3.91
e	420.	30.		3.25
f	750.	30.	7.9	4.02
g	460.	5.		3.38
h	750.	5.	8.7	4.04
i	390.	.5		3.25
k	80.	.5	7.6	2.64

The isentropic efficiency of the turbines is assumed to be 0.90. The equilibrium reaction of combustion is as follows:

$$CH_4 + 2O_2 \rightarrow CO_2 + 2H_2O + (55\ MJ/kgCH_4)$$
$$(16)\quad (64)\quad (44)\quad (36)$$

The fuel-air ratio based on the stoichiometric coefficients is 64/16=4, whereas the enthalpy rise during combustion per kg of combustion products is 55/(1+4)=11 MJ/kg. The enthalpy rise after combustion per kg of the steam-CO_2 mixture is:

cd	3.91-0.86=3.05 MJ/kg
ef	4.02-3.25=0.77
gh	4.04-3.38=0.66
total	4.48

The required relation of the total flow rate to combustion products is 11/4,48=2.46. According to molecular mass of carbon dioxide, its flow rate share is 44/(80·2.46)=0.224. For CH_4 it is 0.081 and for O_2 0.325.

In many textbooks, the specific power consumption to produce oxygen is around 0.3 kW.h/kg O_2. Perhaps in some new devices it may be less, but for the sake of reliability, this figure shall be used. Thus, per kg of the working fluid, it is

$$D_{O_2}=0.3\cdot 3600\cdot 0.325=350\text{kJ/kg}$$

CO_2 compression from subatmospheric pressure up to some tens of bars is needed to cross the saturation line. In Fig. 6.10, the process is presented as a straight horizontal line but in fact many "teeth" exist. Power consumption is approximately* evaluated by the product of the temperature, entropy change and mass share divided by the compression efficiency, i.e.,

$$D_{CO_2}=(0.224)(288)(5.95\text{-}4.95)/0.7=92\text{ kJ/kg}$$

The cycle efficiency (see the state points in Fig. 6.4) is

$$\frac{de+fg+hi-D_{O_2}-D_{CO_2}}{cd+ef+gh}=\frac{660+640+790-350-92}{4480}=0.37$$

This figure is comparable to that of ordinary plants with stacks.

The numerator above, 1648 kJ/kg, is the specific power per kg of flow. Hence, for a power plant of 1.6 GW, a flow rate of 1000 kg/s is needed. In this case, the CO_2 flow equals 224 kg/s. Assuming a velocity of liquid CO_2 of only 1m/s and a density of 1250 kg/m^3, the required pipe cross-section is 224/1250 (1)=0.18 m^2. The pipe diameter is 0.48 m.

Looking at this figure we may conclude that disposing of the liquid carbon dioxide even with long distances is a mild problem. It could be used to

---

* Here we neglect the deviation of CO_2 from an ideal gas which gives over-estimation of the compression work. In more rigorous calculations of CDICWEG cycles it was taken into account. (see Figure 6.20)

neutralize some basins containing alkalies. The global storage of CO_2 certainly might be the ocean, as is clear in Fig. 6.8. The use of carbon dioxide in the ocean to enhance photosynthesis is worth attention. The absorption capacity of the ocean many times exceeds the ability of mankind to burn fuel. Even though the solubility of CO_2 in low pressure warm water is quite small, in deep cold water it is very high (see Fig. 6.1).

After the first guess mentioned above, the computer simulation of this cycle has been made by K.N. Zvagolsky and V.A. Gavrilenko. In the Fig. 6.5, the computer-made flow-sheet is presented with a fixed and changed temperature indication. The internal efficiencies of turbines were assumed different: 0.8; 0.85 and 0.9 for HP, MP and LP turbines. The steam properties were calculated simultaneously.

As the result, the total efficiency variation and SUSEX criterion versus cycle temperatures are presented. The dotted line on Fig. 6.6 presents efficiency with the maximum cycle temperature variation. The two other lines reflect intermediate temperature variation with fixed t_{max}=1300°C.

When t_{max}=Td=750°C as in the previous guess, the efficiency equals 0.35 instead 0.37 before. It is a reasonable agreement. It is clear that by 1300°C, the efficiency might be 0,43 despite CO_2 compression and O_2 production.

The net exergy coefficient K_ε here is very high. It is the inherent property of energy conversion with high density of energy currents. Therefore, a deviation of η_Σ from η_ε is small and SUSEX on Fig. 6.7 does not differ from inversed exergy efficiency.

This cycle is not new. Many of its important features have appeared in earlier papers; one of these is by Kuzminsky and another by Christianovich (see Zysin). From 1892-1900, a combustion chamber with steam injection was developed by P.D. Kuzminsky. In his test rig, the steam of high pressure water flowing in the wall of the chamber was used. In the early sixties, a group headed by S.A. Christianovich developed a power process which our cycle resembles closely. The steam injection into the combustion gases, three combustion chambers and turbines for high, medium and low pressures and a maximum temperature of 750°C, are the same. Their experimental work on combustion chambers and other components is a source of useful information. Unfortunately, after initial success, these investigations were cancelled.

Two features set the GOOSTWEG cycle apart from that developed by Christianovich. They are the complete elimination of nitrogen and the disposal of the liquefied effluents in the deeps, features which are necessary in view of the *greenhouse effect* which at the time of Christianovich project was forgotten.

The GOOSTWEG cycle also requires the use of a big air splitting machine which, for a 1 GW unit, is many times greater than any existing on the market today. However, this technology is well developed so that the required equipment can be constructed.

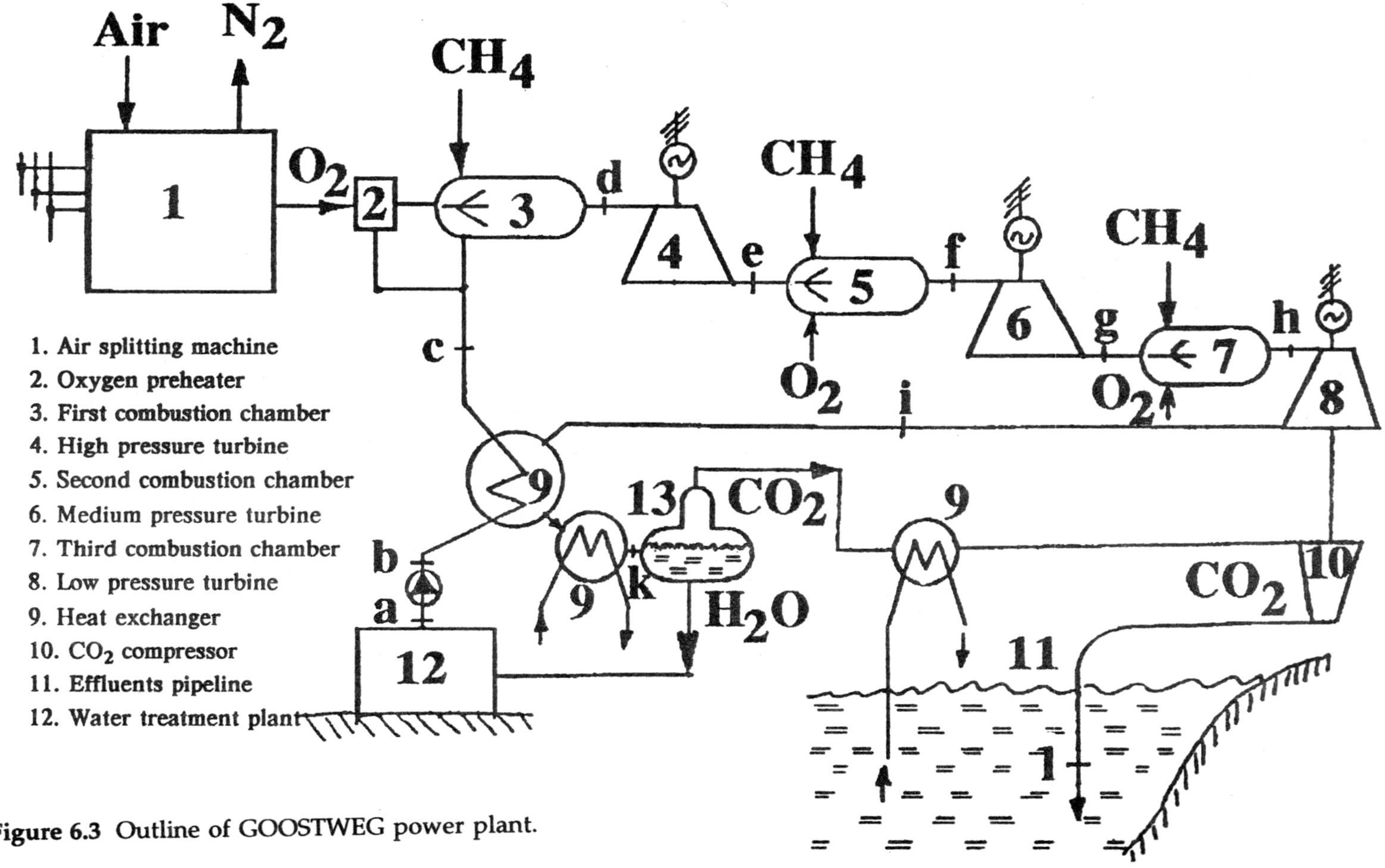

Figure 6.3 Outline of GOOSTWEG power plant.

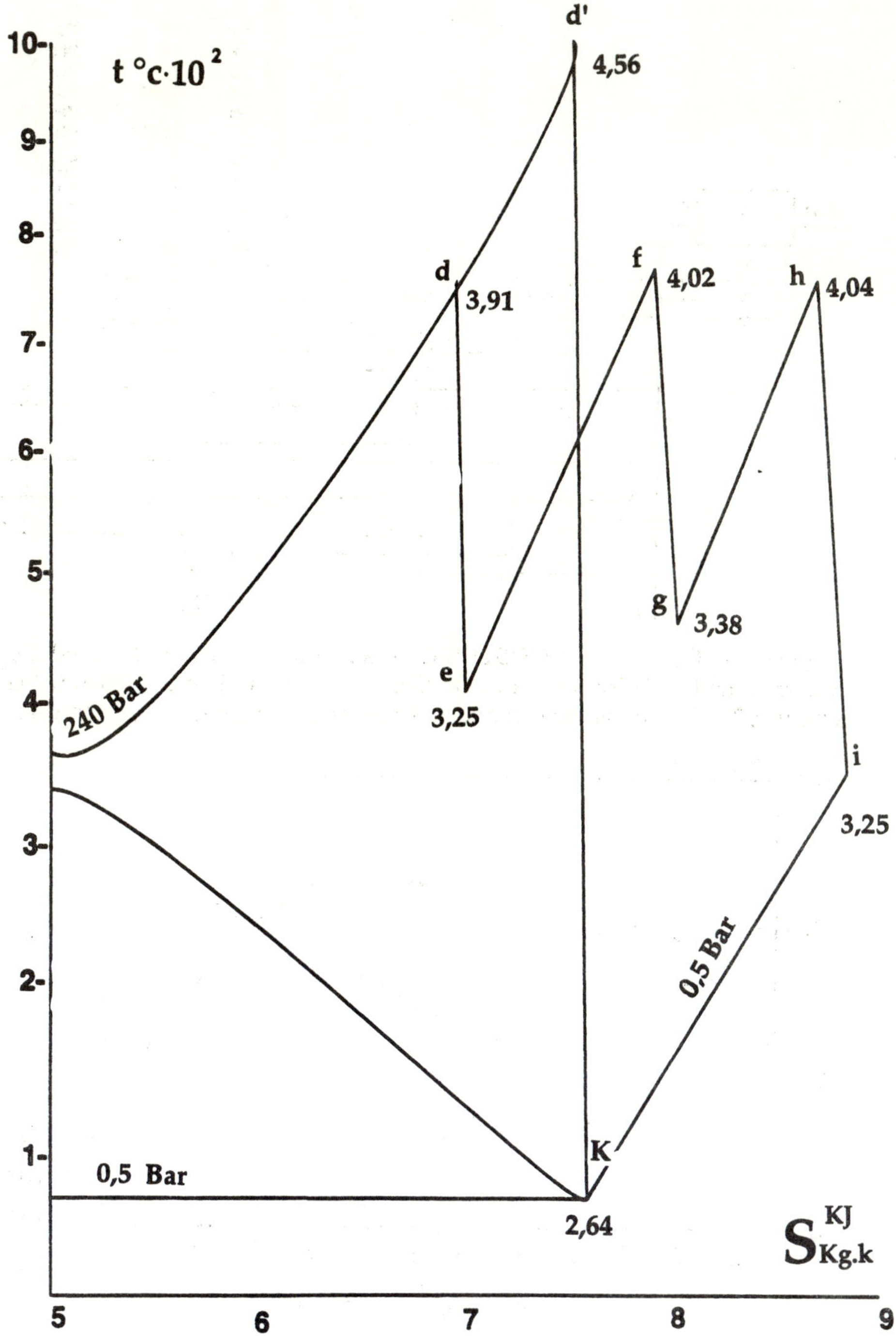

t °c·10²
10
9
8
7
6
5
4
3
2
1
d'
4,56
d
3,91
f
4,02
h
4,04
g
3,38
e
3,25
240 Bar
i
3,25
0,5 Bar
K
0,5 Bar
2,64
S KJ/Kg.k
5
6
7
8
9

Figure 6.4

Ta:	80	Pa:	0.5
Tb:	85	Pb:	240
Tc:	?	Pc:	240
Td:	1300	Pd:	240
Te:	?	Pe:	30
Tf:	1000	Pf:	30
Te:	?	Pg:	5
Th:	750	Ph:	5
Ti:	?	Pi:	0.5
Tk:	?	Pk:	0.5
Tl:	15	Pl:	50

(4) : 80%
(6) : 85%
(8) : 90%

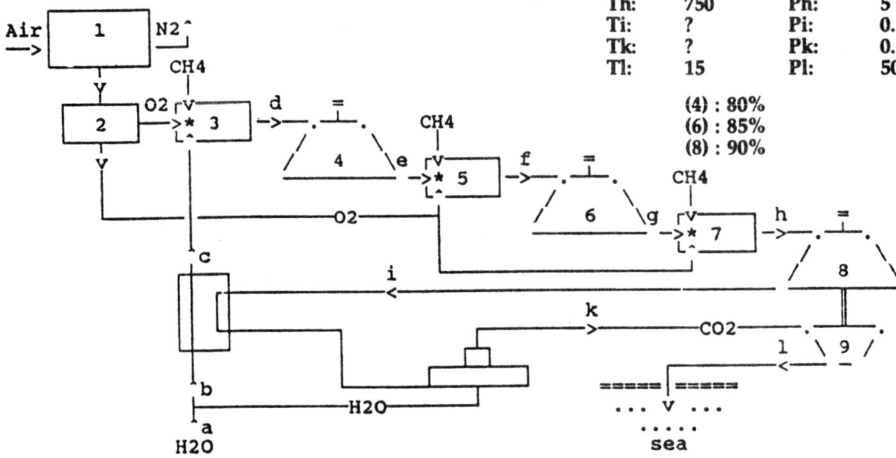

Figure 6.5 Outline of GOOSTWEG power plant (Gas or Oil fired Oxygen Steam Turbine Without Exhaust Gases). 1–air splitting, 2–water/oxygen mixture, 3,5,7–combustion chambers, 4,6,8–steam turbines, 9–CO_2 compressor.

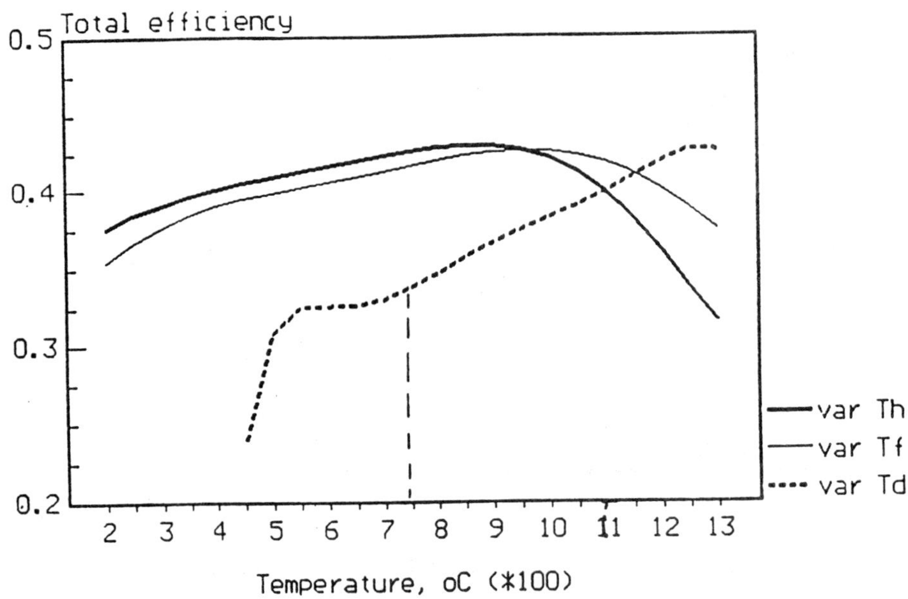

Figure 6.6 Total efficiency of GOOSTWEG cycle versus cycle temperatures.

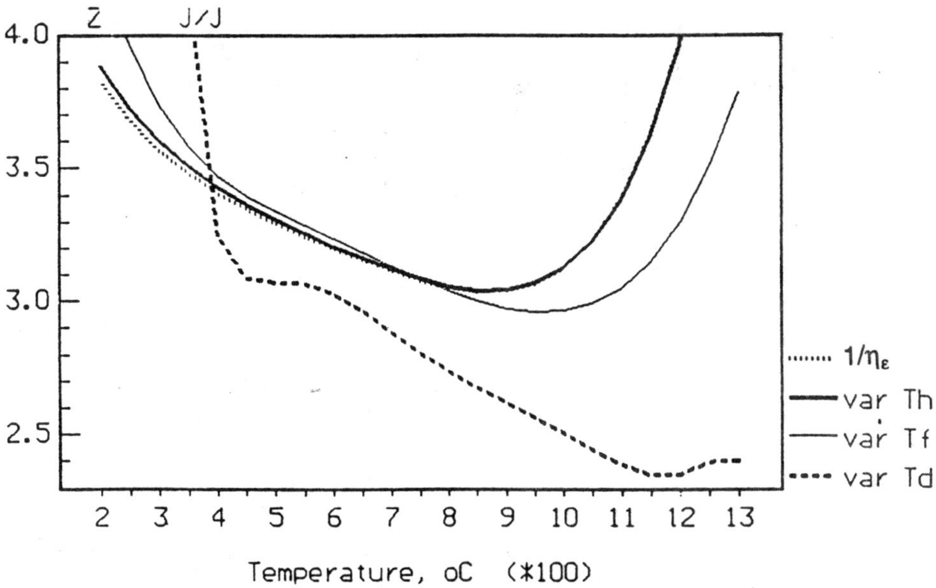

Figure 6.7 Main exergonomic criterion $Z=1/\eta_{\varepsilon}+1/K_{\varepsilon}$ versus cycle temperatures. Point line near solid one reflects small influence of invested exergy.

The COMHDWEG Cycle

The problem of the ecologically benign use of coal and low grade fuels for power production will last for a long time. An attempt to solve this problem in the wake of the previous approach is presented here.

In view of some of the particular properties of this cycle, we must restrict ourselves to power plants near a sea coast. There are many such plants in Europe, Japan, Israel and other countries.

The scheme of this cycle with its proposed MHD power plant is presented in Fig. 6.8. Air from the atmosphere enters the splitting plant where it is separated into oxygen and nitrogen. The latter is released back into the atmosphere without any chemical reactions and is, therefore, harmless. The pure oxygen which is compressed up to 100 bars via a stainless steel heat exchanger enters the combustion chamber.

The fuel consists of coal powder particles mixed in the strong brine of sea water, the waste of desalination. In effect, it is the slurry of fuel and seed (i.e., the potassium chloride of sea salt). The combustion of this slurry in pure oxygen, preheated up to 1100 °K, creates a very high temperature of nearly 3800°K.

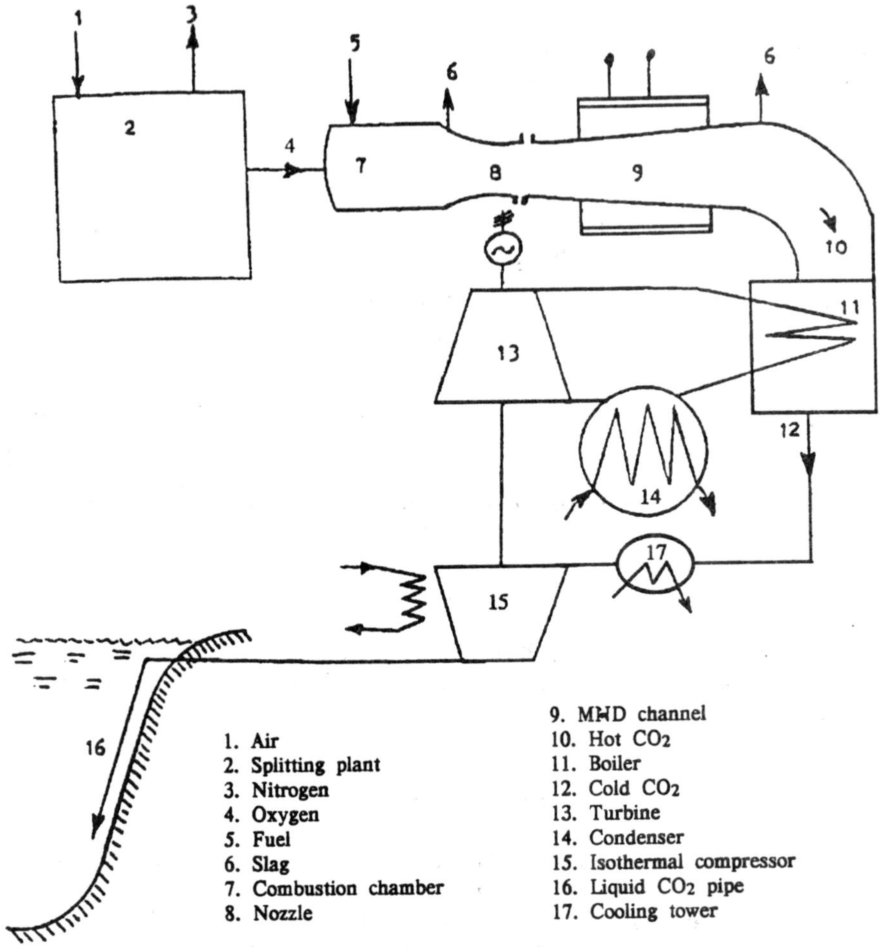

Figure 6.8 COMHD-WEG power plant scheme. 1-air; 2-splitting plant; 3-nitrogen; 4-oxygen; 5-fuel; 6-slag; 7-combustion chamber; 8-nozzle; 9-MHD channel; 10-hot CO_2; 11-boiler; 12-cold CO_2; 13-turbine; 14-condenser; 15-isothermal compressor; 16-liquid CO_2 pipe.

At the exit of the combustion chamber, some part of the liquid slag is released while the other part continues to flow in the MHD channel to protect the walls and electrodes. After isentropic expansion in a nozzle, the combustion gases with their high conductivity enter the MHD channel and the conducting gas (plasma) flowing in a transverse magnetic field create an electromotive force and electrical current in the circuit which consists of electrodes, gas stream and external load resistance (see Yantovskii and Tolmach). It should be noted that in view of the very large pressure ratio required (100 to 1), a disk shaped MHD generator with circular magnetic lines might be recommended (see Fig. 6.9).

After the MHD channel, the residual slag has to be absorbed, along with the rest of the seed, while hot carbon dioxide via a diffuser enters the boiler of an ordinary steam cycle. Some of its heat is used to preheat the oxygen. It is assumed that all the seed was absorbed by the liquid slag and after cooling is consumed in the construction industry or elsewhere. No seed return is needed since for coastal plants the strong brine in the fuel is not worth saving.

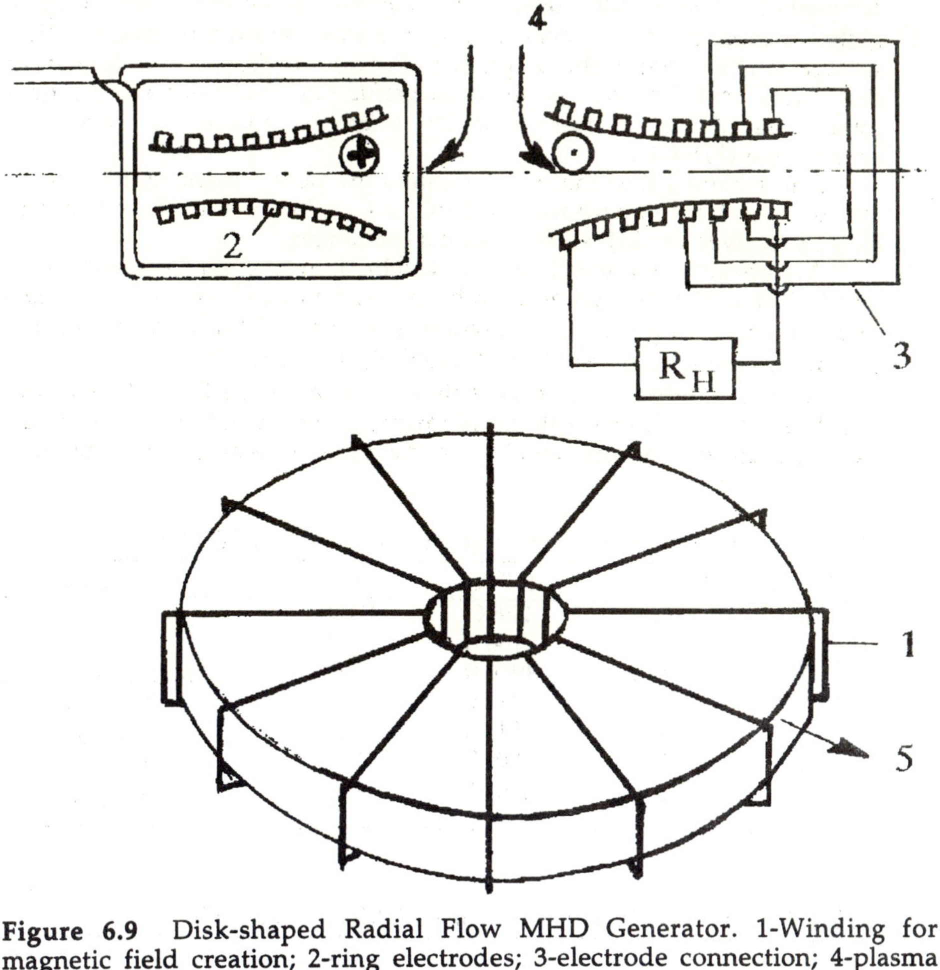

Figure 6.9 Disk-shaped Radial Flow MHD Generator. 1-Winding for magnetic field creation; 2-ring electrodes; 3-electrode connection; 4-plasma from combustion chamber; 5-gas outflow.

In the boiler and cooling tower, the flow of carbon dioxide is cooled down to the sea water temperature at which point the primary new process of the cycle begins, i.e., the liquefaction of the carbon dioxide by means of a simple isothermal compression. In fact there are no pure isotherms in real processes so that such processes actually consist of many adiabatic compressions and intercooling by means of sea water. A temperature of about 15°C was selected.

Compressed wet CO_2 vapor is directed through a steel pipe to the depths of the sea (500 m) where it can be released without damage to the atmosphere. The world's oceans can absorb all the man-made CO_2. In the global carbon cycle, the oceans play the most important role of dissolved storage as is evident in the diagram of Figure 6.10 from B. Giovannini and D. Pain. Note here that the dissolved inorganic (36700 Gt) exceeds 3-6 times the total mass of carbon fuels and 50 times the amount of carbon in the atmosphere (725 Gt).

It is assumed that the oxygen enters the power plant with a pressure of 100 bars and a temperature of 20°C. Its enthalpy I=0. At point one, the oxygen receives I=1 MJ/kg from a heat exchanger.

According to the diagrams, the combustion heat for Donets-basin coal is I_r=32 MJ/kg. The stoichiometric ratio is L=2.6 in kg O_2 per kg of C (Fig. 6.11). The enthalpy rise per 1 kg of mixture is 32/3.6=8.9 MJ/kg CO_2 and the heat loss in the combustion chamber (5%)=.45 MJ/kg CO_2.

All the state points of this cycle are presented in Fig. 6.11 and in Table 6.2. As in any preliminary thermodynamic cycle estimations, some reasonable assumptions on the efficiencies of the main processes are needed. They appear in Table 6.3.

Table 6.2 Thermodynamic state points for the GOMHDWEG cycle.

Point	I (MJ/kg)	t (°K)	P (bars)	Content
0	0.0	288.0	100.0	O_2
1	1.0	1100.0	100.0	O_2
2	9.45	3900.0	100.0	CO_2+seed+dissociate
3	5.4	2740.0	1.0	CO_2 gas
4	4.3	2540.0	1.0	CO_2 gas
5	0.0	288.0	1.0	CO_2 vapor
6	0.0	288.0	50.0	CO_2 liquid

Table 6.3 Process efficiencies for the COMHDWEG cycle.

	(m)	(w)	(b)
Combustion chamber	0.95		
MHD isentropic including excitation	0.7	0.6	0.8
Steam bottoming	0.4	0.35	0.45
Isothermal CO_2 compressor	0.7	0.6	0.8
Calculated cycle efficiency	0.385	0.31	0.46

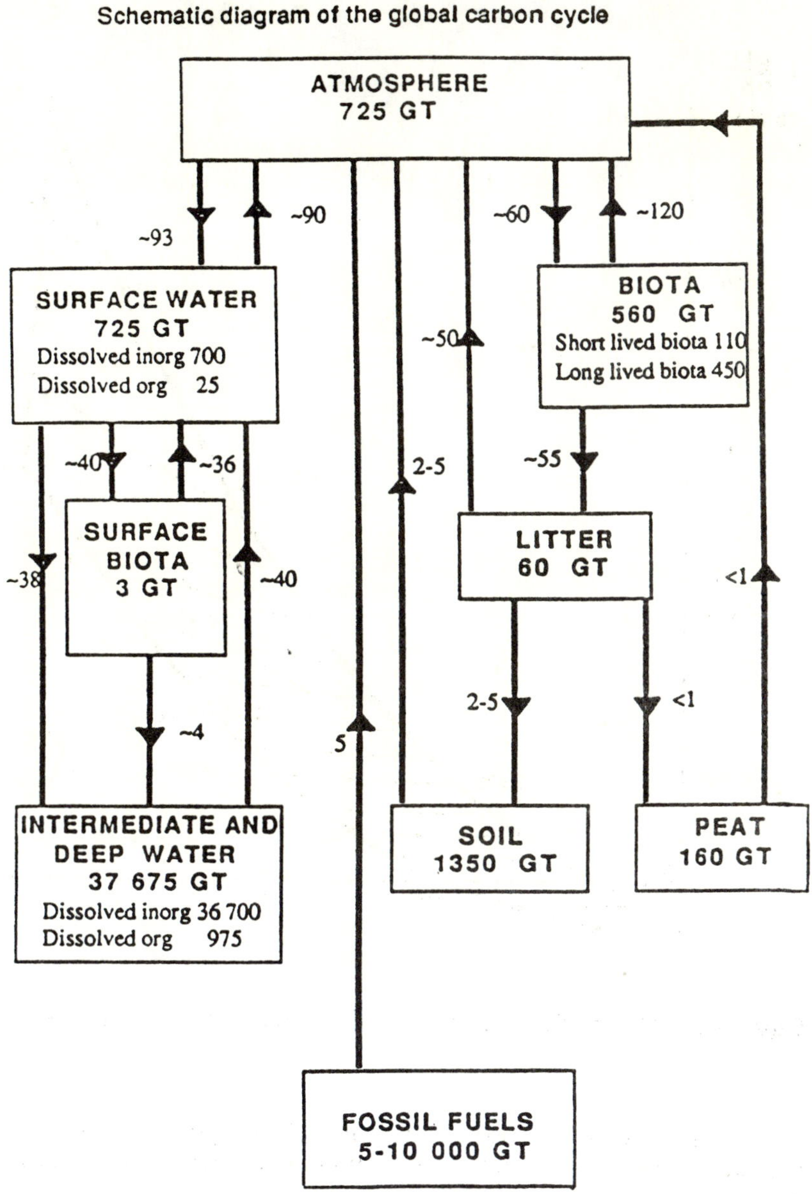

Figure 6.10 Schematic diagram of the global carbon cycle. Quantities are in gigatons (1 GT=10^9 tons) of carbon.

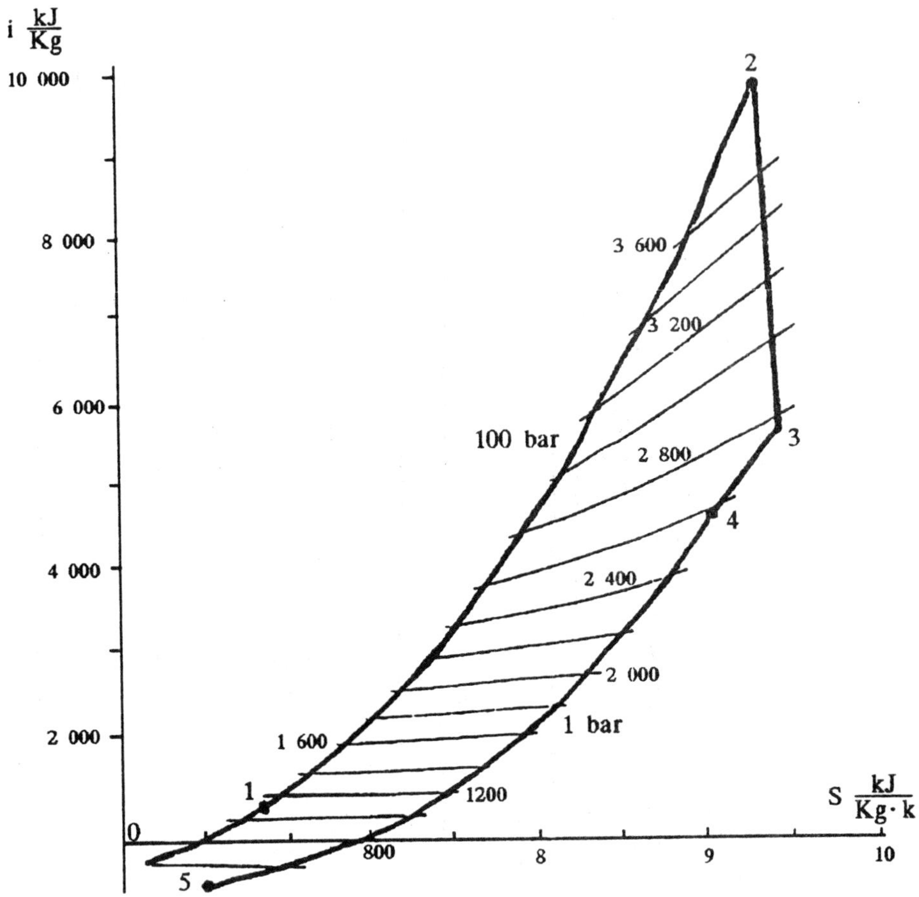

Figure 6.11 Thermodynamic cycle of the coal combustion in pure oxygen.

In this table, the moderate (m), worst (w) and best (b) cases are presented. Even the worse case is suitable in view of the cleanliness of the plant.

The power consumption to get 1nm3 of O_2 is assumed to be 0.4 kWh which translates into 1MJ/kg O_2 or .72 MJ/kg CO_2. The power to compress CO_2 is equal to 288 (5.95-4.95)/0.7=.41 MJ/kg in the moderate case.

The efficiency calculation for this case is as follows:

$$\text{cycle efficiency}=[(4.05)(.7)+(4.3)(.4)-.41-.72]/8.9=.385$$

The numerator here is the specific power per 1 kg CO_2 and is equal to 3.42 MJ/kg. Hence, for a 1 GW plant, the gas flow rate is:

$$\frac{1000 \ \text{MJ}/\text{s}}{3.12 \ \text{MJ}/\text{kg}} = 320 \ \text{kg}/\text{sec}$$

This rather small flow rate is evident because of the lack of nitrogen ballast. At the end of the isothermal compression to 15°C and 50 bars and with a density of CO_2 equal to 143 kg/m^3, its velocity in a pipe having a 0.5 m diameter is quite reasonable, i.e., 13 m/s.

The liquid CO2 velocity, in the same pipe at its exit in the sea depth, is 1.6 m/s. It is quite clear that transferring all the release from a single large 1 GW MHD unit by means of an ordinary steel pipe half a meter in diameter poses no problem.

The COGSTWEG Cycle

This cycle is based on coal gasification. There exist many such processes (see Schilling et al.) and almost all of them use steam and oxygen. For our purposes, the cleanest and most well developed process has to be chosen. Three examples of such processes are the Lurgi, Texaco and Bigas processes. In the Texaco process, the pressure is not more than 18 to 35 bars. Hence the produced fuel gas has to be compressed up to 240 bars. There are no exhaust gases which is very important for the power plant cycle.

The plant outline is presented in Fig. 6.12. The new elements are a coal gasifier (14) and ash ponds (11). The gas of 14 is much less caloric than methane, but this is of no importance because of the regulation of steam flow into the combustion chamber.

In the particular case of alkaline ash, this plant might be very clean, providing the means of neutralizing the ash by means of the carbonic acid of dissolved CO_2.

If the gasifier efficiency equals 0.75, then the total power plant efficiency related to the coal input is (.75)(.39)=0.29. This efficiency could obviously be increased by increasing the temperature at the turbine inlet; but even with the rather low efficiency of 0.29, this cycle could be attractive for coal consuming countries.

The COGSTWEG cycle offers a lot of possibilities for producing extremely clean, nature conserving power. Typical emissions in the USA, for coal-fired plants per GJ of energy delivered, is as follows: CO_2=240 kg, SO_2=90 g, NO_x=90 g and suspended particles equal to 6 g. Among 19 projects of clean coal-fired power plants, including the Integrated Gasification Combined Cycle (IGCC), not a single radical solution has appeared. Current reports indicate that the removal of CO_2 from combustion gases would raise the cost of electricity by 35 to 80% with the final rate of CO_2 still an uncertainty. The COGSTWEG cycle seems to be helpful here.

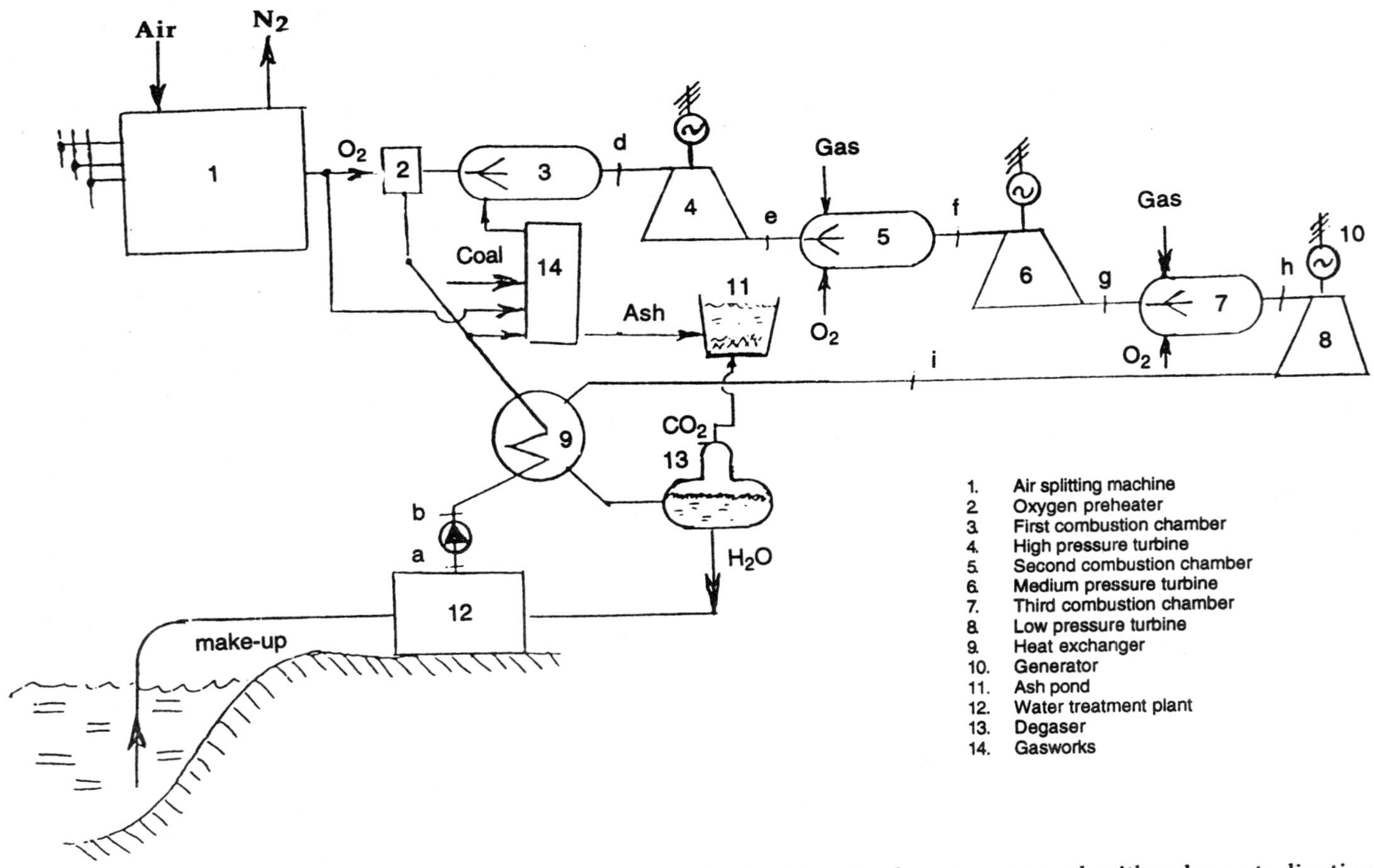

Figure 6.12 Outline of coal-oxygen steam turbine power plant without exhaust gases and with ash neutralization by carbon acid.

The SOFT Cycle

The layout for the SOFT power plant is presented in Fig. 6.13. Air enters the splitting machine (1) where it is divided into nitrogen and oxygen. The nitrogen is released to the air. In the mixer (2), oxygen flow is injected into hot high pressure steam. This mixture and the fuel enter the first combustion chamber 3 where the temperature of the steam and CO_2 as combustion products increases to a maximum.

Steam expands in the high pressure turbine (4) which drives an electrical generator. After the first expansion,the fuel and oxygen are injected and burned in chamber 5 and then again in chamber 7. After the low pressure turbine (8), the mixture of CO_2 and steam enters the recuperator (9) where partial condensation takes place. However, the CO_2 is still a vapor. The total condensation of steam and the dissolving of CO_2 occurs in the jet condenser (10) where pumped water from the pond (14) is used.

The pond (14) is the most important part of the plant. It is filled by fresh water with photosynthesis making algae chlorella. The aim of the pond is to absorb solar energy in order to convert dissolved CO_2 into organic matter which is then used as fuel for the power plant. The oxygen from the reduced CO_2 is released to the atmosphere. Under steady-state conditions, the production and consumption of oxygen are equal.

These *renewables* depend mainly upon the exergy flow concentration. As a rule, the natural exergy current density is low, on the order of $100W/m^2$ (on the average). However, some rare exceptions exist. One is the hot jets of *blacksmokers* in the ocean depths where at a temperature of 350°C the exergy current reaches $1Gw/m^2$ (see Chapter 5).

One of the best concentrators of natural exergy is photosynthesis. In the global energy flow diagram in Fig. 6.14 (Häefele), one can see something resembling a law of natural energy flow, i.e., the less the flow the higher the concentration. Large flows with hot humid air (comprising 97.5% of total solar energy input) have very low current densities, such as the one mentioned above ($0.1kW/m^2$). Warm water and ocean currents are nearly 3% of the total solar energy input and have a current density of $1MW/m^2$. If we consider the flow of water with microalgae, e.g., green river water, with the suspension of only 1% of organic matter, a velocity of 2 m/s and a combustion heat of 30 MJ/kg, the current density is:

$$10^3(2)(30)(10^6)(10^{-2})=0.6 \text{ GW/m}^2$$

Hence, in the case of biomass which is only 0.04% of the total solar energy input, the current density might be a million times greater than in moving air or in solar light, keeping in mind the solar constant of 1.3 kW/m^2 or 0.6 kW/m^2 behind atmosphere.

The water treatment plant (12) also acts as a refinery, dividing the mixture of water and algae into pure high quality water and organic matter which is then used as fuel. The concentration of energy takes place at first in photosynthesis itself, then in the flow of water with organic material and finally in the plant itself (12) when water is separated from the fuel. Here a slurry of organic matter in water may be used.

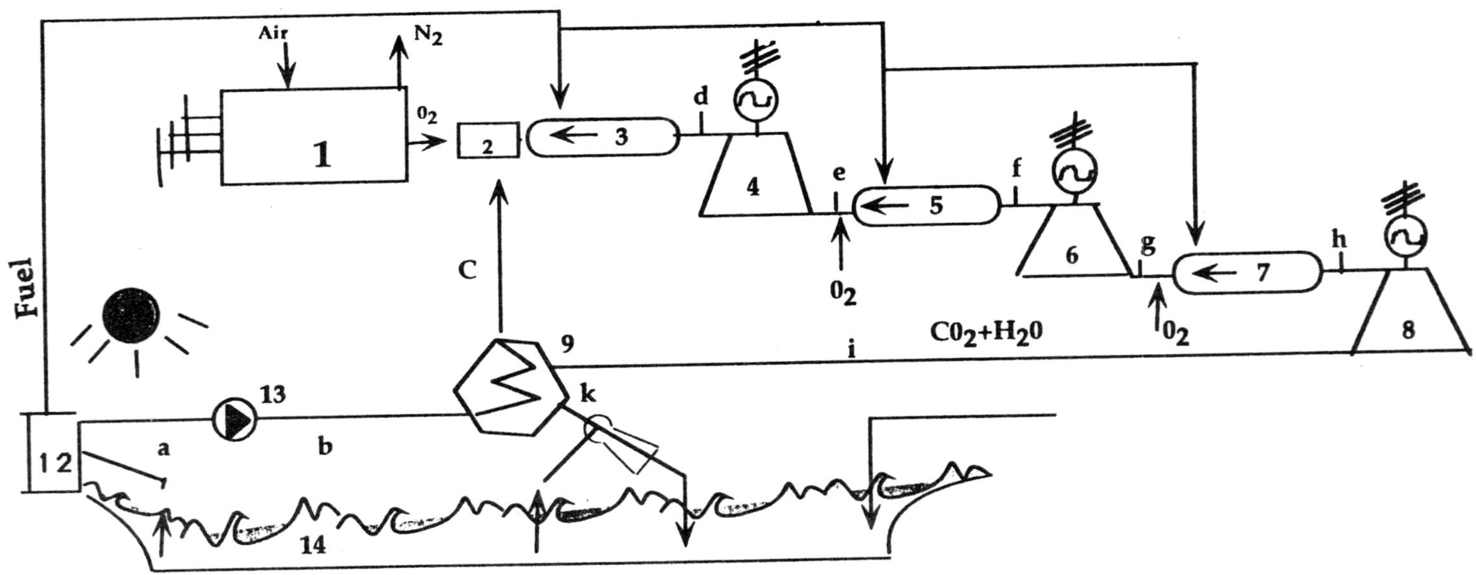

Figure 6.13 Outline of SOFT power plant. 1) Oxygen plant, 2) Steam-oxygen mixer, 3) First combustion chamber, 4) HP turbine, 5) Second combustion chamber, 6) MP turbine, 7) Third combustion chamber, 8) LP turbine, 9) Recuperator, 10) Jet condenser, 11) Make-up water, 12) Fuel and water plant, 13) Feed water pump, 14) Photosynthesis solar pond.

The primary energy of this cycle is solar falling on the ponds surface. The converted energy is in the form of electricity from the generators. The losses are heat losses from equipment and low grade heat losses including evaporation from the pond's surface. Therefore, make-up water (11) is needed.

The cycle itself is closed with respect to the oxygen and carbon and nitrogen is not a component of the cycle. The only need is exergy from the Sun; and according to the Second Law, the entropy flow from the pond exceeds the entropy coming from the falling sun beams.

The best place for this type of plant is in a densely populated belt between 20 and 40 degrees latitude (e.g., Greece, Italy, Spain, Florida, Texas, China, Uzbekistan, Turkey, Israel, etc.). Being clean, such plants might be situated close to consumers. The average solar energy current here equals 250 W/m^2.

The crucial question of the whole project is the efficiency of the photosynthesis. This problem has been considered in detail in Semenov's book, especially in the comprehensive review of 332 references by K. Zamaraev and V. Parmon which appears on pages 43 through 82. The limiting figure given for this efficiency is 0.15. It is irrelevant to the Carnot factor and depends upon the quantum character of light absorption. All the quanta with an energy less than the threshold one are useless and even those absorbed transfer only a part of their energy. The most reliable data for real calculations can be found in the experimental work of some of the research groups at the University of Moscow (Alexeev et al.). In order to avoid mistakes, some of the pertinent text is translated literally on the following page, i.e.,

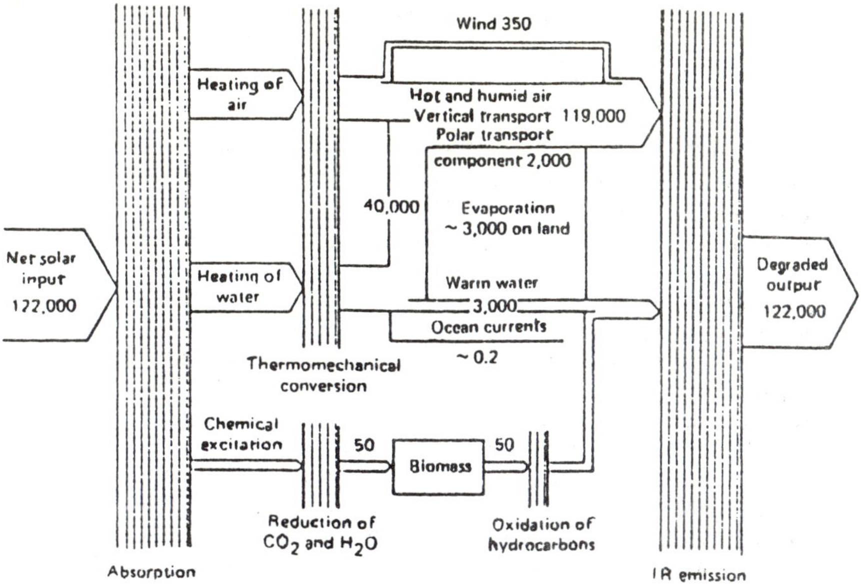

Figure 6.14 Global energy flows.

Since 1980, the development of the solar energy converter referred to as the "biosolar" has been done by the physics faculty, chemistry faculty and some other faculties. Based on this work, the microalgae are planted over a large surface with subsequent methane production. The mineralized elements from the methane tanks are reused by the algae. Carbon is used in the form of CO_2 *and delivered in the final product as* CH_3. *After burning, the* CO_2 *is restored. Therefore, the global change of* CO_2 *in the atmosphere is absent. The combustion of fossil fuel, on the other hand, contributes to the greenhouse effect due to the increase of* CO_2 *in the atmosphere.*

The surfaces which are to be covered by photosynthesis plants depend on energy intensity. An efficiency of up to 20% of active radiation is achieved in the laboratory. Certainly, this high figure is not likely under natural conditions. In fact, experience with the massive production of microalgae shows that the real efficiency is between 8 and 10%.

If the average radiation reaches 240 W/m^2*, 40 g of dry substance can be produced in a day from 1 m*^{2}*, i.e., 1 mtce of methane could be produced from a surface of 70 km*^{2}*.*
An outline of "biosolar" is presented in Figure 6.15 for a mixture of the microalgae Chlorella Pyrenoidosa and Ankistrodesmus Braunii which mutually use the metabolism products.

From the pump, the slurry of biomass in water enters the methane tank where a gas 80% CH_4*, 16%* CO_2*, 2%* H_2 *and less than 2% of other constituents is produced.*

Benemann considered many inorganic reducing agents (H_2, H_2S, NH_3 etc.) and stated "only photosynthetic processes using water for CO_2 reduction are likely applicable for CO_2 mitigation. A laboratory-scale system, involving algae growth, digestion to methane and recycle of the nutrients was successfully demonstrated (Golueke and Oswald, 1959)...algae biomass cultivation in open ponds could be relatively inexpensive...it is possible to produce microalgae in large-scale outdoor ponds at both high productivities and at relatively cost." The two different productivities were assumed in calculations, "about 110 and 220 metric tons/a year, corresponding to about 5 and 10% solar energy conversion efficiencies, for a site with good climate."

Brown and Ziegler gave more information to it. For algae pond "saline water can be used if necessary. Microalgae are so unique among photosynthetic organisms in that they can achieve extremely high productivites at salinities as high as twice that of sea-water. In contrast to the limited available arable land for forestry, much desert land is available globally for microalgae farming. Microalgae require only 140-200 lb of water per pound of carbon fixed in open ponds."

Ability to use salt water and even brines is of great value to the concept of MHD generator without exhaust gases, (see COMHDWEG cycle in this Chapter.) The organic fuel as a slurry of algae in the brine forms electrically conductive combustion products in the MHD channel without need of special seeding by potassium due to its high concentration in salty water.

The both concepts of SOFT cycle and COMHDWEG might be brought together. One can use the organic matter of the algae as a fuel of MHD generator, where it is to be burnt in pure oxygen. The combination is of value in case of sea water algae pond with some salt content in the fuel. It would play the role of ionizing seed to get high electrical conductivity of hot combustion gases. The same salt is harmful for the turbine use.

Alexejev et al. have presented experimental data of real photosynthesis efficiency between 8 and 10%. "If average radiation reaches 240 W/m^2, 40 g of dry substance can be produced in a day from 1 m^2, i.e. 1 mtce of methane could be produced from a surface of 70 km^2."

The last figure might be compared to Benemann data: "Land area required (500 MW Power Plant) 1 mtC/y CO_2 production, growth pond x 1.2, 6000ha." It means 60 km^2 which is in excellent agreement with mentioned Alexejef figure of 70 km^2. The overall reactions for photosynthesis by ciano-bacteria, microalgae and plants in solar light according to Brown and Zeigler are as follows:

$$2H_2O = O_2 + 4H^+ + 4e^-$$

$$CO_2 + 2H^+ + 2e^- = CH_2O$$

In short:

$$CO_2 + H_2O = CH_2O + O_2 - Q_f$$

Benemann recomended the figure of Q_f=7.5 kkal/g, it means 31.3 MJ/kg CH_2O. It is quite enough to any of internal combusion cycle due to the need of large amount of recirculated inert gas, either steam or carbon dioxide. Therefore our reference equation of combustion reaction and molecular masses are:

$$\underset{30}{CH_2O} + \underset{32}{O_2} = \underset{44}{CO_2} + \underset{18}{H_2O} + 31.3\ MJ/kg\ CH_2O$$

The primary concern of calculations are thermodynamic properties of flue gases acting as working substance with changing contents of CO_2 and H_2O. For pure water and carbon dioxide we make use recent data (Altunin, Rivkin) where all the needed figures are given.

The data for mixture we got by mass fraction addition of enthalpies and entropies. In the iteration process on every step the flue gases composition and properties were calculated simultaneously. Computer indicated the number of iterations, as usual it was 5-15.

The temperature t_c after recuperator was selected so low as to avoid "pinch effect." In other words there exists sufficient temperature drop between heating and heated flows, not less than 10 C.

Here is the list of assumed efficiencies of processes:

Oxygen compression	0.8	Recuperator	0.9
Feedwater pump	0.6	First turbine	0.8
Second turbine	0.85	Third turbine	0.9

Efficiency calculations example (see Table 6.6)

$$\eta = \frac{493 + 658 + 449 - (87 + 54 + 18)}{2318 + 575 + 639} = 0.408.$$

The primary information on the results is presented in Tables 6.4-6.6. The only change in conditions is turbine inlet temperature $t_d = t_l = t_h = 550$, 650, 750 C. Mass fractions, turbine power, state points are calculated simultaneously.

Table 6.4

mCH_2O 0.103	mO_2 0.110	mCO_2 0.152	mH_2O 0.062	$(1\text{-}mCO_2\text{-}mH_2O)$ 0.786
	P bar	T °C	H kJ/kg	S kJ/(kg&K)
a	1	80	335	
b	240	85	376	
c	240	219	948	
d	240	550	3108	6.0260
e	30	239	2713	
f	30	550	3274	7.0639
g	5	217	2737	
h	5	550	3254	8.0422
i	1	346	2899	
k	1	95	2400	
l	60	35		

Efficiency = 0.338.
Turbine 1: Hde=367
Turbine 2: Hfg=519
Turbine 3: Hhi=355
O_2 production: D_{O_2}=79
O_2 compression: L_{O_2}=50

Water pump: L_{H_2O}=19
Chamber 1: Hed=2160
Chamber 2: Hef=562
Chamber 3: Hgh=517

Note, that temperatures after expansion in points e, g and i are indicated for isoentropic process, while enthalpies in the same points are in accordance to mentioned turbine efficiecies with some entropy rise.

Table 6.5

mCH_2O 0.108	mO_2 0.115	mCO_2 0.159	mH_2O 0.065	$(1\text{-}mCO_2\text{-}mH_2O)$ 0.776
	P bar	T °C	H kJ/kg	S kJ/(kg*K)
a	1	80	335	
b	240	85	376	
c	240	263	1150	
d	240	650	3386	6.3548
e	30	306	2910	
f	30	650	3480	7.3117
g	5	280	2865	
h	5	650	3445	8.2657
i	1	434	3053	
k	1	95	2388	
l	60	35		

Efficiency = 0.375

Turbine 1: Hde=440 — Water pump: L_{H_2O}=19

Turbine 2: Hfg=591 — Chamber 1: Hed=2236

Turbine 3: Hhi=392 — Chamber 2: Hef=570

O_2 production: D_{O_2}=83 — Chamber 3: Hgh=579

O_2 compression: L_{O_2}=52

Table 6.6

mCH_2O 0.113	mO_2 0.120	mCO_2 0.165	mH_2O 0.068	$(1\text{-}mCO_2\text{-}mH_2O)$ 0.767
	P bar	T °C	H kJ/kg	S kJ/(kg*K)
a	1	80	335	
b	240	85	376	
c	240	299	1328	
d	240	750	3646	6.6469
e	30	394	3110	
f	30	750	3685	7.5146
g	5	348	2997	
h	5	750	3636	8.4392
i	1	506	3187	
k	1	95	2376	
l	60	35		

Efficiency = 0.408

Turbine 1: Hde=493 — Water pump: L_{H_2O}=18

Turbine 2: Hfg=658 — Chamber 1: Hed=2318

Turbine 3: Hhi=449 — Chamber 2: Hef=575

O_2 production: D_{O_2}=87 — Chamber 3: Hgh=639

O_2 compression: L_{O_2}=54

There is much to be done in view of optimization of the cycle including distribution of temperatures and pressures over the turbines. However still presented efficiencies 0.338-0.408 might be satisfactory. By 40% of cycle efficiency and 10% of photosynthesis on the total efficiency of solar energy into power conversion is 4%. It is much in excess of that of salt-gradient solar pond power (2%). By the average insolation of 250 W/m^2 and total efficiency 0.04 the power per m^2 equals 10 W. For a demonstration power plant of 10 MW the surface of the pond is 1 km^2. For a utility-scale power plant of 1 GW the dimensions of the pond 10x10 km are enough. It is not much that exceeds the dimensions of a circulation water cooling pond of ordinary fuel-fired or nuclear power plant.

In comparison to the Salt Gradient Solar Pond, the SOFT cycle seems to be better because it needs two times less land and wind is not as detrimental for the SOFT pond as for the salt gradient stability of the latter.

The depth of the pond can not be too great since solar beams could not then penetrate to the bottom of the zone. As is well known, the infrared part of light is absorbed within the upper few centimeters whereas the intensive part of ultraviolet light can reach several meters deep into clear water.

Now, in view of the seasonal variation of the power demand, the SOFT plant is indeed soft since the fuel produced can be stored and the turbine power easily regulated. On a seasonal basis, some fuel is stored during the sunny summer days and CO_2 after the winter demand. If one wishes to store fuel for a year for a 1 GW plant, 2 mt of fuel are needed. If the content of the fuel in water is 1%, the water volume must be 0.2 km^3 and for a surface area of 100 km^2, the depth 2m.

The net energy payback time for the SOFT plant seems to be very short because solar energy is not falling on silicon high energy consuming material but on the surface of water in a pond which is much less energy intensive in its construction.

The equipment for such a SOFT power plant belongs to proven technology, the turbines are just of-the-shelf.

Having looked at the controversial statements of investigators of repositories for CO_2 (deep ocean, aquifers, etc.) one can see a lot of questions still remained in the problem of carbon dioxide sequestering. This problem is solved in SOFT cycle automatically, the store-house for combustion born CO_2 is just the pond with algae. We do not need to wait for the end of investigations in oceanology or geology in our attempt to build a demonstration plant.

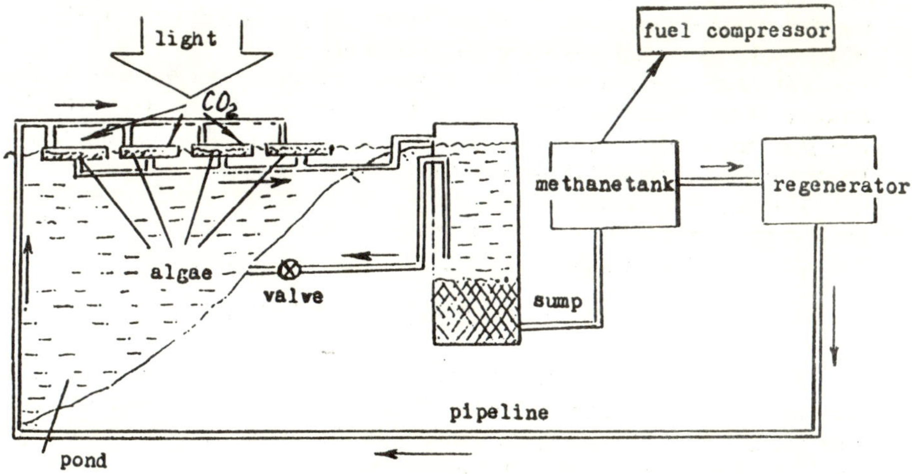

Figure 6.15 "Biosolar" outline.

Proposed Carbon Dioxide Cycle

We learned already that good properties of CO_2 as a working substance for power cycles are known since 1940, when it was proposed by D.P. Hochstein for external combustion and then for nuclear power.

However en route to that goal a semiclosed scheme had been presented, which in effect is the first, to our knowledge, internal combustion carbon dioxide scheme, see Figure 6.16.

Here the oxygen plant (ASU, air separation unit) encompassed by the dotted line 1 is linked to power plant within line 2.

Liquid water born in combustion is separated and extracted in the points 20 and 21. Carbon dioxide by means of pump 18 after heating is directed to the combustion chamber 8 along with fuel and oxygen from the compressor 7. The total flow of recirculated and combustion born CO_2/steam is expanded in the turbine 9. An excess CO_2 in liquid form is released in the point 19.

For the sake of simplification in the present chapter we consider the scheme with a separate oxygen plant, not integrated as on Figure 6.16 to power plant. The oxygen plants at the moment are well developed mature technology, therefore we wish to let well alone. The presented scheme where optional chemical recuperation version is added, see on Figure 6.17.

The cycle is presented on Figure 6.18 and 6.19-6.20.

We wish to explain the logic of our calculations and simplification of the model by means of numerical example. Here we assume:

– the lack of pressure drops in heat-exchangers and internal pipes, due to easy pressure elevation in liquid CO_2 if needed,

– the lack of heat losses from the surface of heat-exchangers and combustion chambers, due to moderate temperatures compatible with good thermal insulators. However in our rigorous computer simulations of COOPERATE and SOFT cycle all these loses were taken into account.

We have selected moderate maximal temperatures because in this case amount of steam in flue gases is small and the use tables for pure CO_2 does not lead to unadmissible inaccuracy. The tables of Altunin book and well known Plak-Kuprianoff diagrams are used.

The turbine isentropic efficiency was assumed as 0.85, carbon dioxide multistaged compressor as 0.80, oxygen power consumption 0.3 kWh/kg O_2.

The chemical reaction taking place in recuperator 3 is

$$CH_4+CO_2=2\ CO+2H_2+h^* \quad (h^*=247\ 478\ \text{kJ/kmole.})$$

It means the conversion of primary fuel, in our case methane CH_4, into secondary, more energetics fuel, carbon monoxide and hydrogen. A proper catalyst is needed for such conversion. In ammonia industry nickel is widely used for that matter along with alumina.

The upper temperature of gases in our case is strongly restricted by the strength of steel in heat-exchanger on the level 900 C. The completeness of chemical reaction α depends upon temperature and catalyst, we assume in our scouting study only 50% of conversion.

The enthalpy consumed in complete reaction equals 15.5 MJ/kg CH_4, for 50% conversion (α=0.5) we will use 7.75 MJ/kg CH4.

The efficiency calculations need to indicate the boundary of the system along with all the incoming and outgoing flows. In our case the boundary surface is crossing incoming flow of methane and outflows of power and liquid carbon dioxide. There are no stack gases.

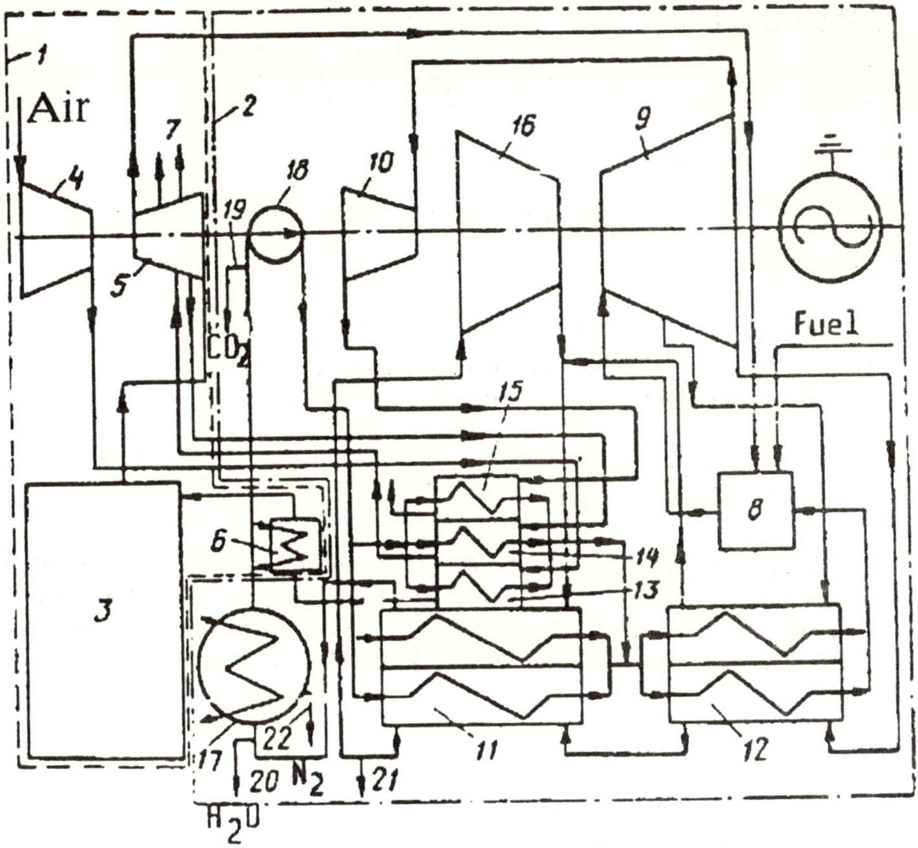

Figure 6.16 Outline of carbon dioxide zero-emissions power plant by V.L. Degtiarev and V.P. Gribivsky of July 28, 1967 [8] 1-Oxygen plant (ASU); 2-Carbon dioxide power plant; 3-Rectification; 4-Air compressor; 5-Oxygen compressor; 6-Air cooler; 7-Compressed oxygen extraction; 8-High pressure combustion chamber; 9-Main turbine; 10-Auxiliary turbine; 11-Combined regenerator; 12-Main regenerator; 13,14,15-Regenerative heater; 16-Compressor; 17-Condensor; 18-Pump; 19-Liquid CO_2 release; 20,21-Water release; 22-Nitrogen release.

For the ordinary efficiency calculations we have to divide net power by the fuel energy flux:

$$\eta = \frac{L_{turb} - L_{compr} - L_{ox} - L_{fuel}}{M_{fuel} \cdot H_{fuel}}. \tag{6.1}$$

For exergy efficiency we need to take into account exergy or produced CO_2 (208 kJ/kg CO_2):

$$\eta_{\varepsilon} = \frac{L_{turb} - L_{compr} - L_{ox} - L_{fuel} + E_x CO_2}{M_{fuel} \cdot E_x CH_4}. \tag{6.2}$$

From evident energy balance we have:

$$M_{tot} \cdot h_{comb} = M_{fuel} \cdot h^{con} = M_{fuel} \cdot (H_{fuel} + \alpha h^*). \tag{6.3}$$

Hence

$$M_{fuel} = M_{tot} \cdot h_{comb} / (H_{fuel} + \alpha h^*). \tag{6.4}$$

The fuel consumption decreased with the increase of αh^*.

After introduction of (6.3) into (6.1) we have:

$$\eta_{cr} = \frac{L'_{turb} - (L'_{compr} + L'_{ox} + L'_{fuel})}{h_{comb}} \cdot \frac{H_{fuel} + \alpha h^*}{H_{fuel}} = \eta \left(1 + \frac{\alpha h^*}{H_{fuel}} \right) \tag{6.5}$$

If H_{fuel}=50MJ/kg CH_4 and αh^*=7.75 MJ/kg CH_4 the second ratio in (6.5) equals 1.155. It means that chemical recuperation can elevate efficiency η in that relation. The unavoided precondition of this increase is our ability to take extra enthalpy from the flow of flue gases after turbine.

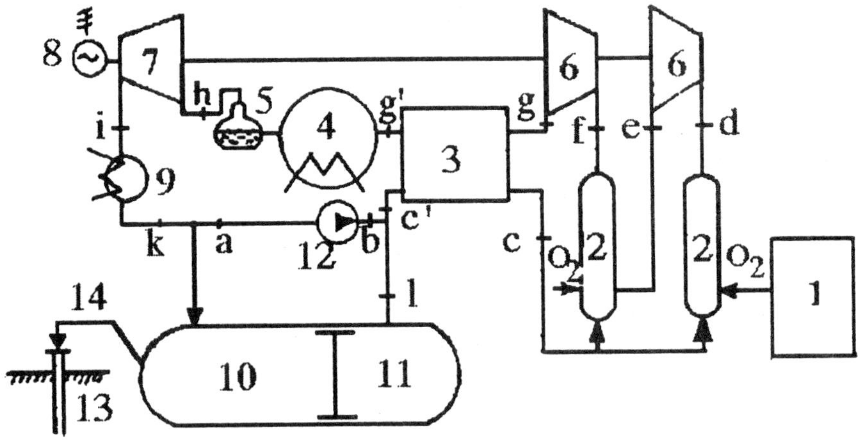

Figure 6.17 Outline of chemical recuperation assisted CDICWEG power cycle. 1-Air separation unit; 2-Combustion chamber; 3-Recuperator with catalyst; 4-Cooling tower; 5-Water separator; 6-Turbine; 7-Intercooled compressor; 8-Generator; 9-CO_2 condensor; 10-CO_2 tank; 11-Fuel gas tank; 12-CO_2 pump; 13-Depleted well; 14-Liquid CO_2 pipe.

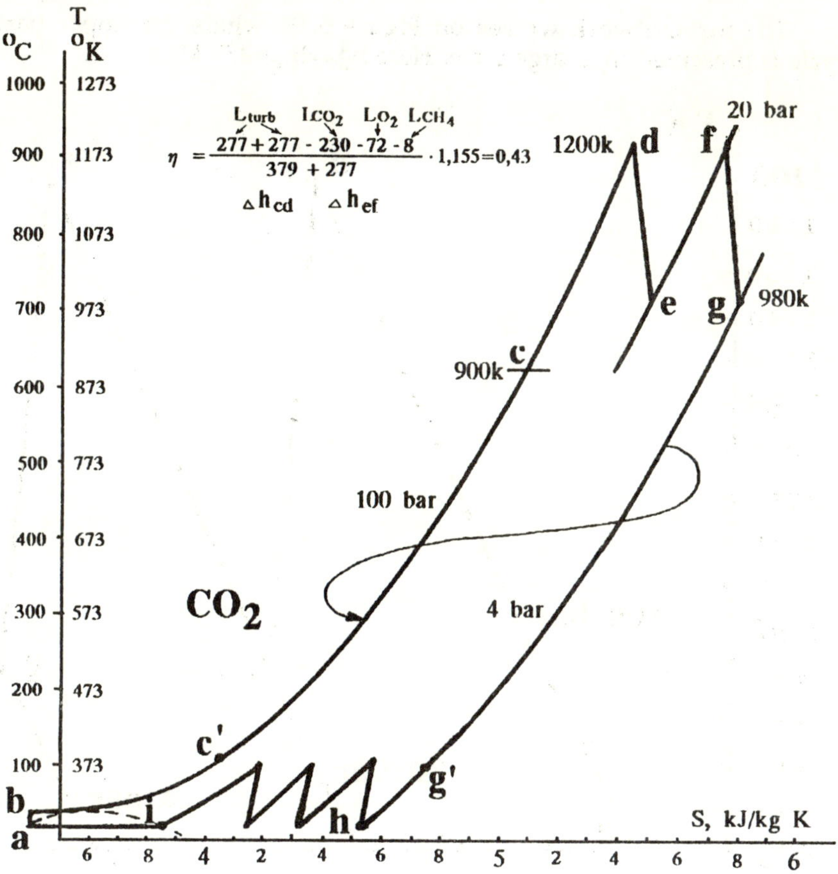

Figure 6.18 **T - S diagram of the chemical recuperation assisted cycle.**

For the first example we have selected maximal pressure 100 bar and minimal 4 bar with the equal expansion ratio 5 in any of the two turbines, see cycle diagram on Figure 6.18.

The mass fractions of the total flow are as follows:

$$m_{CH_4} = M_{CH_4} / M_{tot} = h_{comb} / (H_{CH_4} + \alpha h^*) = \frac{379 + 277}{57750} = 0.0114.$$

Assuming stoicheometric combustion, the oxygen fraction is

$$m_{O_2} = 4m_{CH_4} = 0.0456, m_{CO_2} = 0.0314, m_{H_2O} = 0.0257.$$

Note here large excess of recirculated CO_2 over the combustion born fraction (0.943/0.0314=30). The latter is to be released from the plant in liquid form.

The turbine work we see on Figure 6.19, where the upper part of the cycle is presented in a large scale. Here $h_{de}=h_{fg}=277$ kJ/kg.

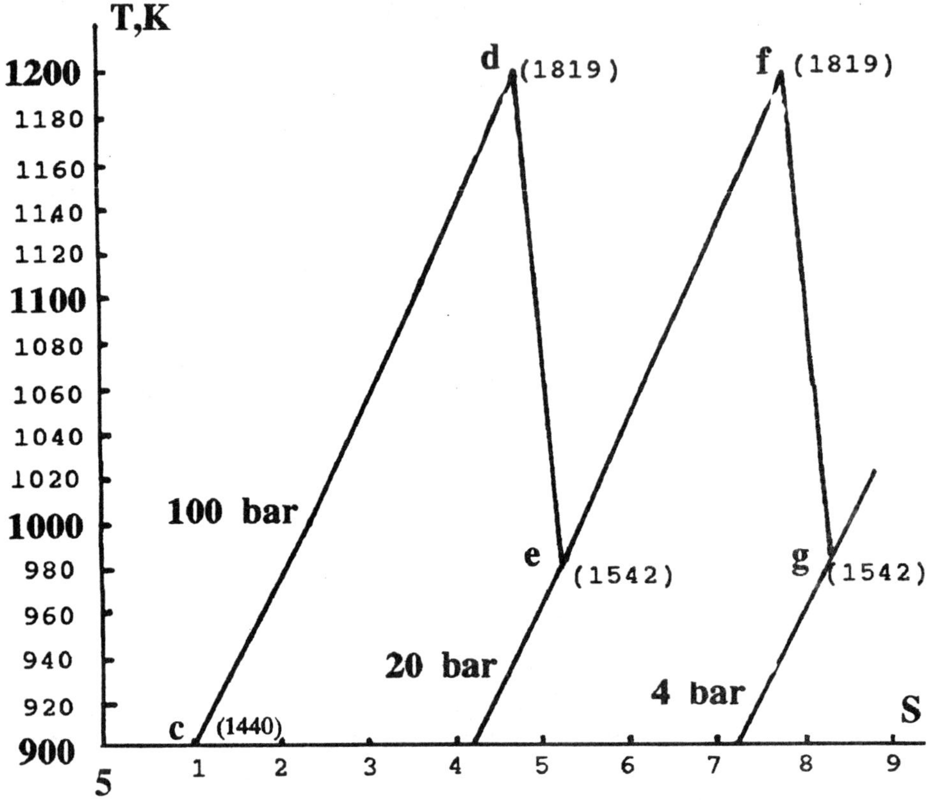

Figure 6.19 Upper part of the cycle on T-S diagram.

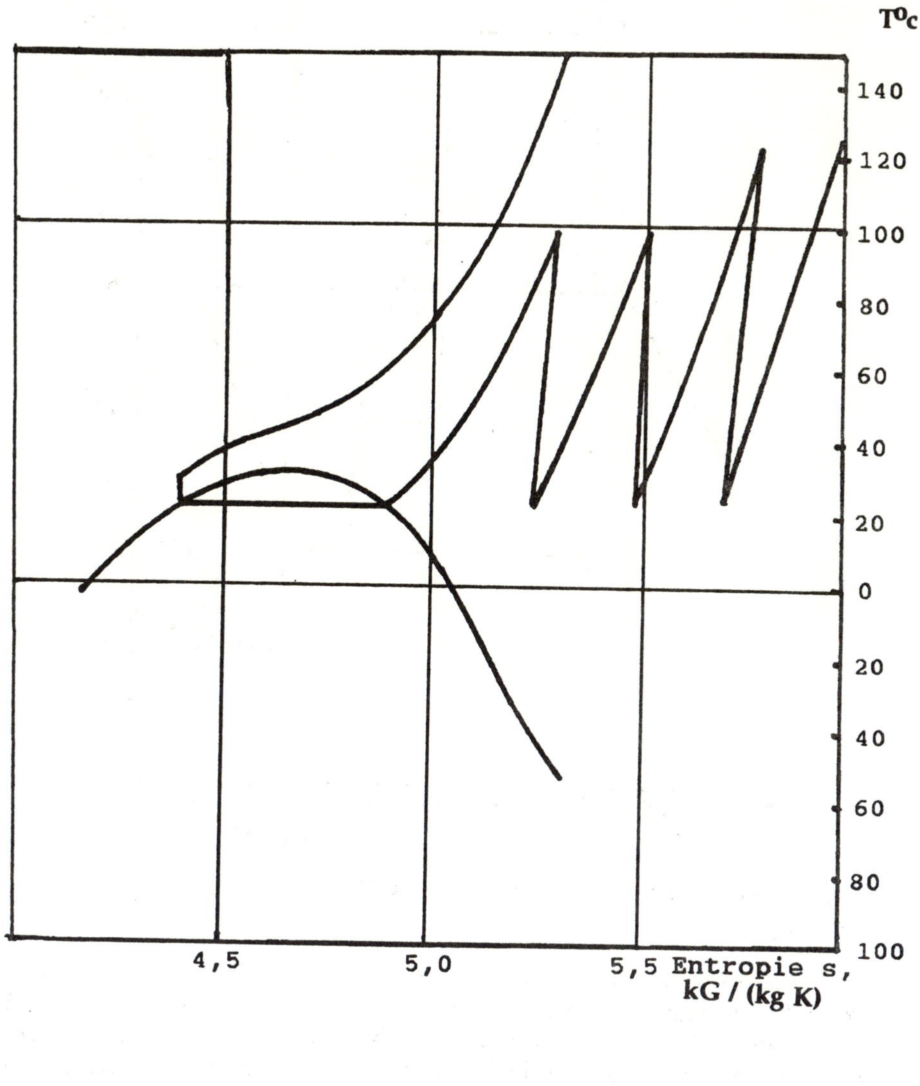

Figure 6.20a The cycle bottom part (CO_2 compression and liquefaction) on T-S diagram.

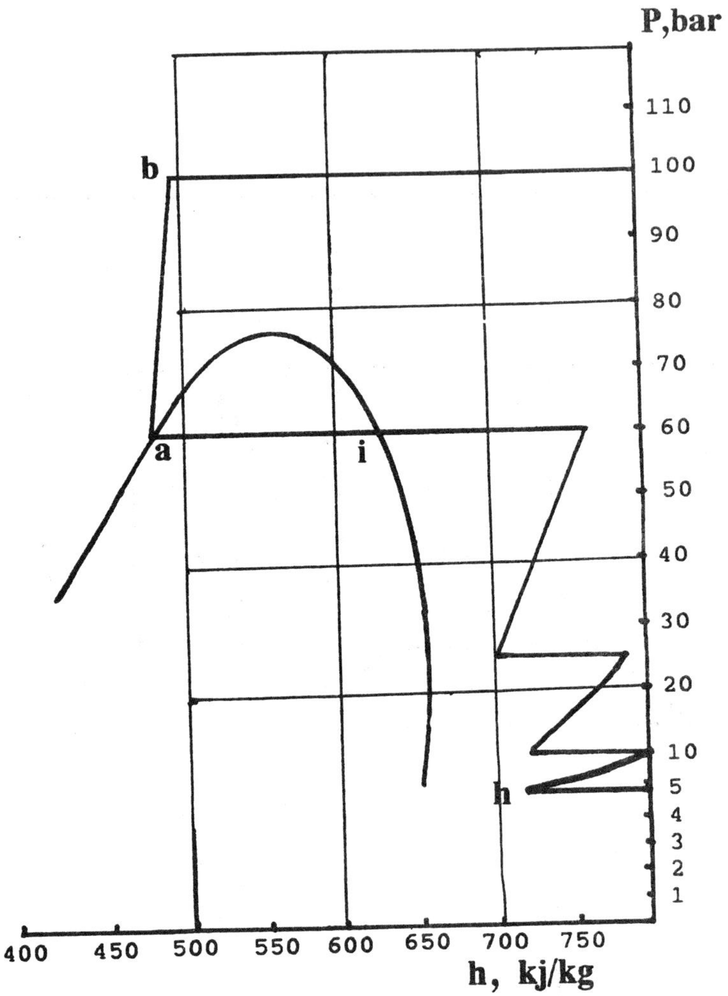

Figure 6.20b Bottom part of the cycle (CO_2 compression and liquefaction) on p-h diagram.

The compression of CO_2 in an intercooled three-stage compressor is presented on Figure 6.20a in T-S coordinates and on Figure 6.20b is duplicated in pressure-enthalpy diagram. The cooling substance is just liquid CO_2 as it was proposed by Degiarev in 1967, to decrease the enthalpy needed to take from the flow of gases after turbine.

The three components of compression work are 88, 65 and 62 kJ/kg, pumping of condensed CO_2 consumes 15 kJ/kg total L_{comp}=230 kJ/kg which elevates temperature up to 370 K.

The power for air separation unit as inherent part of the cycle equals 0.3 kWh/kg O_2 or 0.3·3600·0.0456=50 kJ/kg of total flow. For compression of oxygen from 1 bar up to 100 bar in multistaged (isothermal) machine by efficiency of 0.80 we need:

$$\frac{1}{\eta_c} m_{O_2} RT \ln P / P_0 = 22\, kJ / kg.$$

For methane compression from, say, 3 bar in a pipe to 100 bar in multistaged compressor 8 kJ/kg is needed. The cycle efficiency equals:

$$\eta = \frac{277 + 277 - 230 - 50 - 22 - 8}{656} 1.155 = 0.43.$$

Now we need to approve the energy balance in the crucial device of this cycle, the recuperator. To heat compressed recirculated carbon dioxide the enthalpy rise is as follows:

$$m_{CO_2}(h_c - h_{c'} + m_{CH_4} \cdot h^*) = 0.943(640 + 88) = 686.5\, kJ / kg.$$

From the other side we have enthalpy drop:

$$m_{CO_2}(h_g - h_{g'}) + m_{H_2O} h_{H_2O} = 0.974(1542 - 870) + 0.026 \cdot 1250 = 687\, kJ / kg.$$

Here we take into account the excess of steam enthalpy over that of carbon dioxide. Within an inaccuracy of calculations the energy balance in recuperator has no deficit. At high temperature side the temperature drop equals 980-900=80 K, at low temperature the same is 5 K, it is quite enough to get rather small heat-exchanger by the least pressure 4 bar.

The same calculations for fixed temperature T_c=900 K, but increased pressure (240 bar) and temperature (1400 K) give the results:

$$m_{CH_4} = 0.0166;\ m_{O_2} = 0.066;\ m_{CO_2} + = 0.046;\ m_{H_2O} = 0.037;$$

$$\eta = \frac{375 + 400 - 240 - 104 - 12}{584 + 375} \cdot 1.155 = 0.50.$$

The turbine power here is significantly increased in respect to former example of T_d=1200 K, however CO_2 compression power remained nearly the same due to pumping in liquid state. It is a natural property of CO_2 as a good working substance for power cycles.

If we neglect a small difference between exergy of methane and combustion heat the exergy efficiency should be 0.51 due to addition of exergy fraction of liquid CO_2 9.5 kJ/kg.

In case of 1800 K before turbine the efficiency near 60% might be expected, provided T_c=1000÷1100 K.

These calculations confirm that CDICWEG cycle with chemical recuperation may compete with ordinary combined gas-steam power cycles in respect to efficiency being much in advance in reducing emissions.

There are no real turbines for highest mentioned temperatures on-the-shelf. However rather high temperatures are indicated in projects (Sangiovanni et al.):

For the '90s inlet temperature ranging between 1400 C and 1500 C with peaks at 1600 C are predicted, thanks mainly to increasingly sophisticated cooling systems, to the use of integral discs and blades, composite blades (high-strength metal core and ceramic-plated coating).

High figures of turbine inlet temperature are assumed in the Westinghouse gas turbine 501 F namely 2500 F (1371 C) and 3000 F (1649 C), which represents significant growth challenges beyond current technology in the area of blade cooling (Banister et al.).

Especially important for our case study are projects of steam-injection gas turbines, like General Electric I-STIG cycle LM 8000 and the research project of Institute for High Temperature in Moscow. We selected these examples due to high temperature 1360-1600 C, rather high pressures up to many tens of bar and steam injection. All these features are similar to our case, but for one: emissions. Here all kinds of emission take place, carbon dioxide, nitrogen oxide (very large by that temperatures), hydrogen oxide (steam) in large proportion. The latter needs a lot of pure water make-up, nearly 25 kg/s for 100 MW power. It might be prohibitive.

In our case of zero-emission there are no harm of NO_x formation. It should be dissolved in liquid CO_2 and disposed of. No steam exhaust means no loss of water.

Above mentioned examples let us hope for needed turbine construction with a reasonable confidence. Certainly, very hard engineering problems are the recuperator and high pressure combustion chambers in our case.

Other Comparable Schemes

One of the first recent offers to use internal combustion with CO_2 recirculation belongs to G. Lorentzen and J. Pettersen, see Figure 6.21. They have evaluated efficiency, presented T-S diagram and discussed the cycle. Their conclusion was rather pessimistic due to need of big air separation unit and large recuperator.

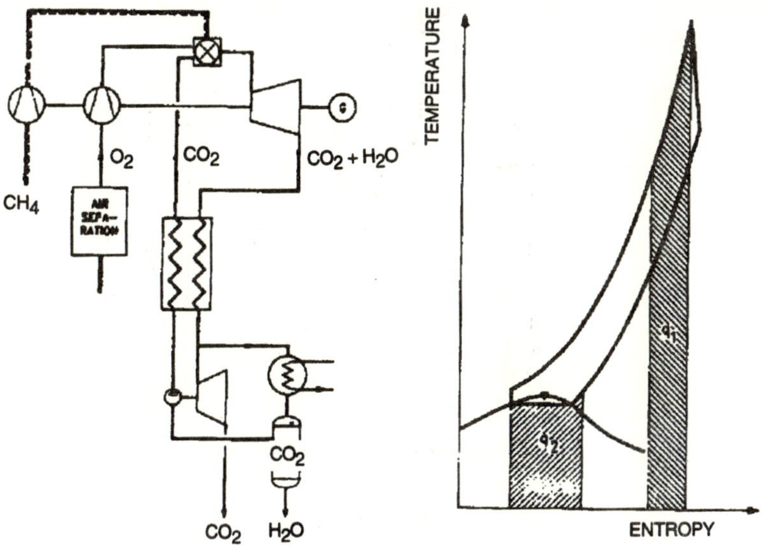

Figure 6.21 Outline and T-S diagram of Lorentzen-Pettersen cycle

The offer of J. de Ruyck which is very close to us, see on Figure 6.22. Here is an oxygen supply, the combustion in the O_2/CO_2 mixture and CO_2 liquefaction. The unique feature is water evaporator similar to the recuperation by means of steam injection into combustion chamber is mentioned already in STIG cycle. However here steam is not exhausted, it is extracted as excess water. In our point of view it is a simple internal combustion cycle on recirculated steam and CO_2 in desired proportion.

Presented efficiency curves versus the specific work indicated very high figures up to 55% by 1200 C and 33% by 860 C. For this zero-emission plant the author concluded that global efficiency exceeding the 50% can be expected in spite of more than 5 points decrease due to the oxygen production.

The next example is the draft of P. van Steenderen, see Figure 6.23. Here all the features of the internal combustion are presented. However the bottoming Rankine cycle is used to avoid some troubles with the recuperation. It is combined cycle. The calculated results for a coal-fired plant with integrated gasifier are presented too.

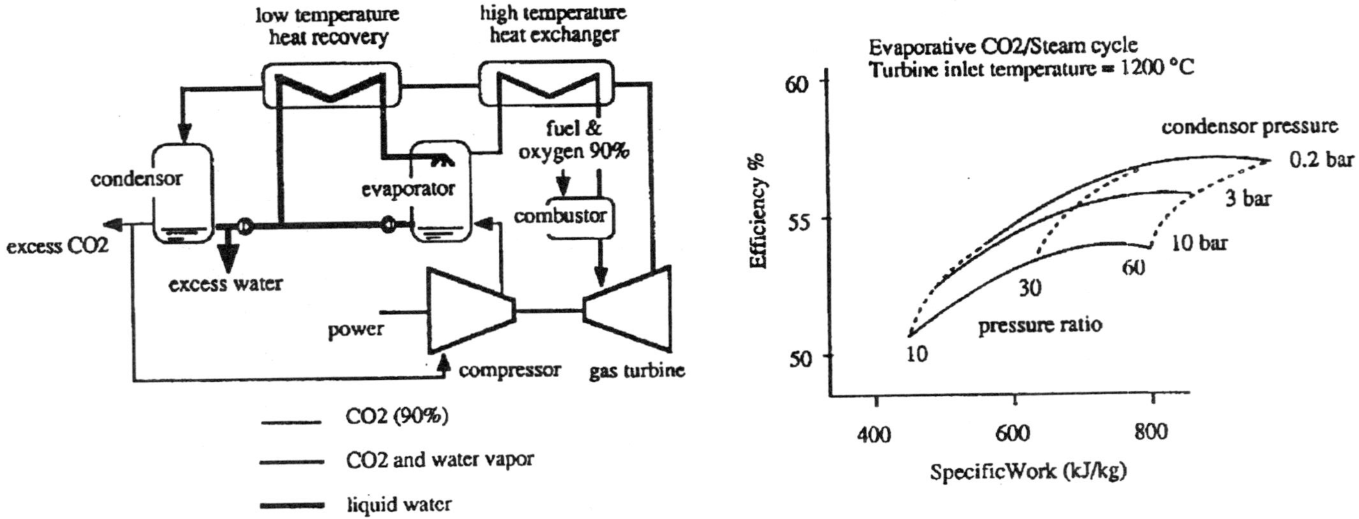

Figure 6.22 Outline and efficiency of zero-emissions De Ruyck cycle. Turbine inlet temperature-1200°C; Turbine adiabatic efficiency-88%; Compressor adiabatic efficiency 85%; Overall pressure loss-6%; Air heater min temperature diff.-100°C; Water heater min. temperature diff.-20°C; Evaporation exit temperature-dewpoint; Condensor inlet temperature-dewpoint.

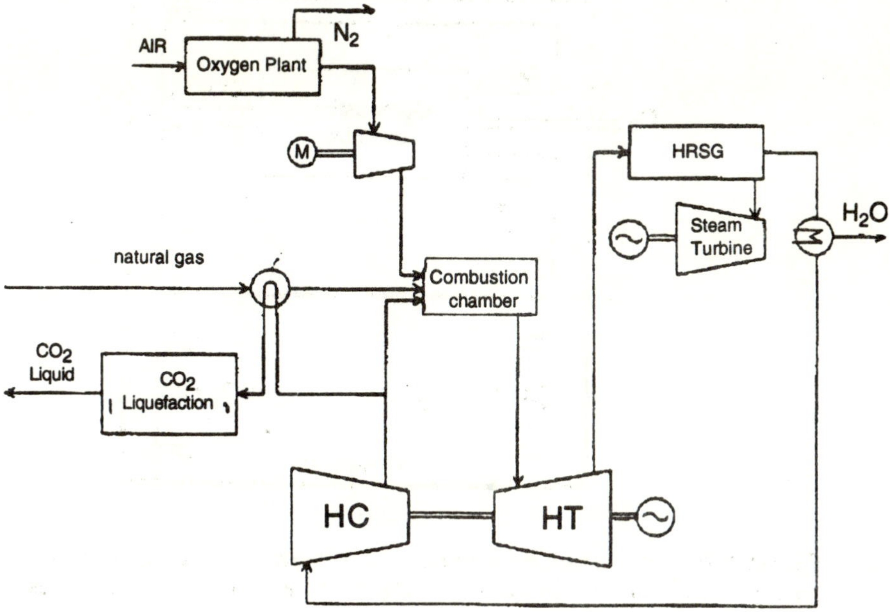

Figure 6.23 Outline of combined gas and steam turbines cycle for natural gas, van Steenderen project.

In this comprehensive study the author concluded that by 20 bar and 1050 C before gas turbine the total efficiency equals 44% and by 11 bar 42.7%.

The paper by T. Holt and E. Lindberg contains calculations on the storing capacity of depleted oil and gas fields for CO_2 disposal of. Two thirds of CO_2 produced by oil and gas combustion might be returned underground, to the same space from where the fuel was extracted. They briefly considered a simple internal combustion power plant with the liquid CO_2 injection for oil recovery enhancement.

The authors focussed on the very important feature of carbon dioxide cycles, to use as a fuel the mixture of extracted hydrocarbon gases and CO_2 injected to enhance extraction. The more CO_2 is going along with HC the less is to be recirculated, see Figure 6.24.

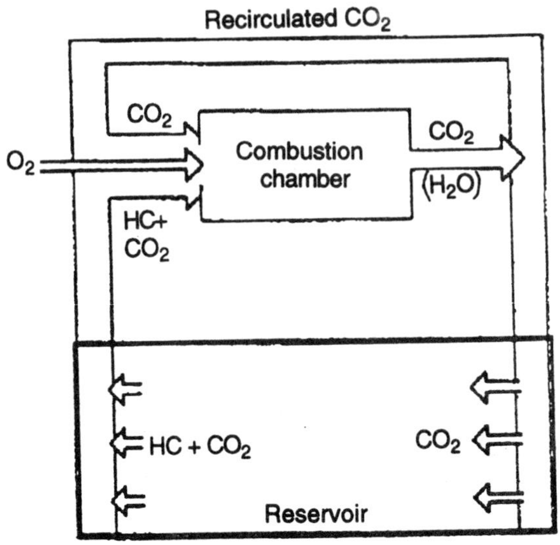

Figure 6.24 The scheme of mass flows between zero-emission power plant with internal combustion in O_2/CO_2 mixture and hydrocarbon reservoir. CO_2 addition to HC in reservoir might be compensated by decrease of recirculation flow.

The last but not the least paper we wish to mention for comparison is that of O. Bolland and S. Saether. To our knowledge it is the best paper on the internal combustion concept.

There the standard combined cycle with emissions was compared to four new cycles. The first pair of them is dealing with the downstream recovery of CO_2 from the flue gases after combustion, it is beyond the scope of our interests. The last two are: a simple internal combustion H_2O cycle, which authors referred to as a Rankine cycle with feed of oxygen, see Figure 6.25, and a combined cycle internal combustion.

In most features the mentioned H_2O cycle is similar to GOOSTWEG scheme, where the three-stage expansion was used and a little less oxygen power consumption assumed.

For the combined cycle the temperatures were not indicated, but it might be as those of Siemens power unit. The results are a little less than that of van Steenderen and ours. For the H_2O simple cycle the graphs of efficiencies with and without the blade cooling are presented, see Figure 6.25. The same is transferred on the Figure 6.26 as solid line and might be compared with our dotted line from our calculations.

Now we can compare all the efficiency evaluations versus cycle temperature on the Figure 6.26. Here we see that we can not expect a significant increase of efficiency in the case of H_2O internal combustion due to large condensation heat of water and modest possible recuperation.

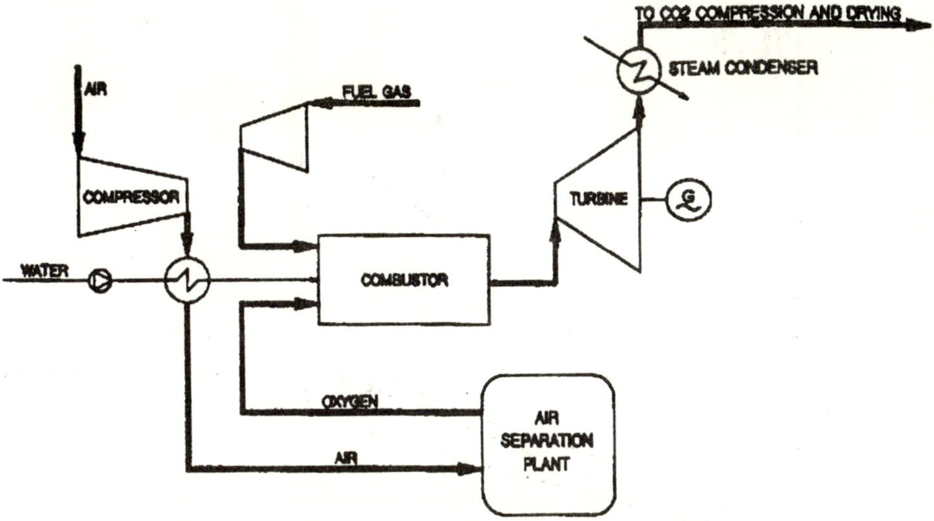

Rankine cycle with feed of oxygen from an air separation plant

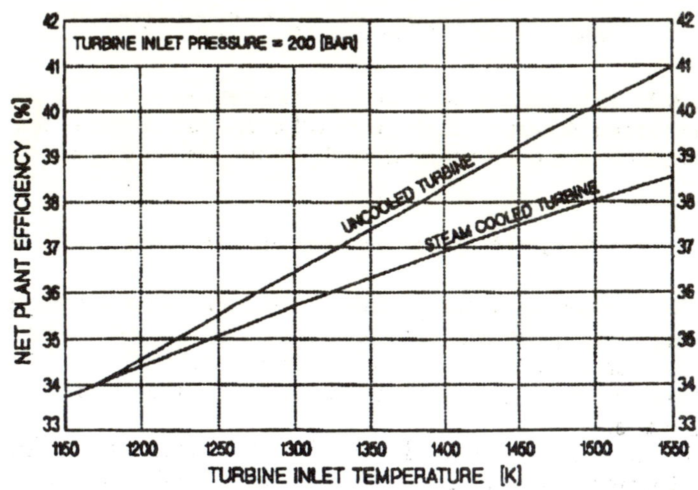

Figure 6.25 Natural gas in O_2/steam mixture combustion zero-emission plant proposed by Bolland and Saether and its efficiency versus turbine inlet temperature.

The simple CO_2 internal combustion cycle seems to be more promising. In spite of unavoidable discrepancy of the database, different assumptions etc. the trend of increase of efficiency up to 60% by the temperature 1800 K is evident. Reasonable power consumption to get oxygen has been taken into account in all papers.

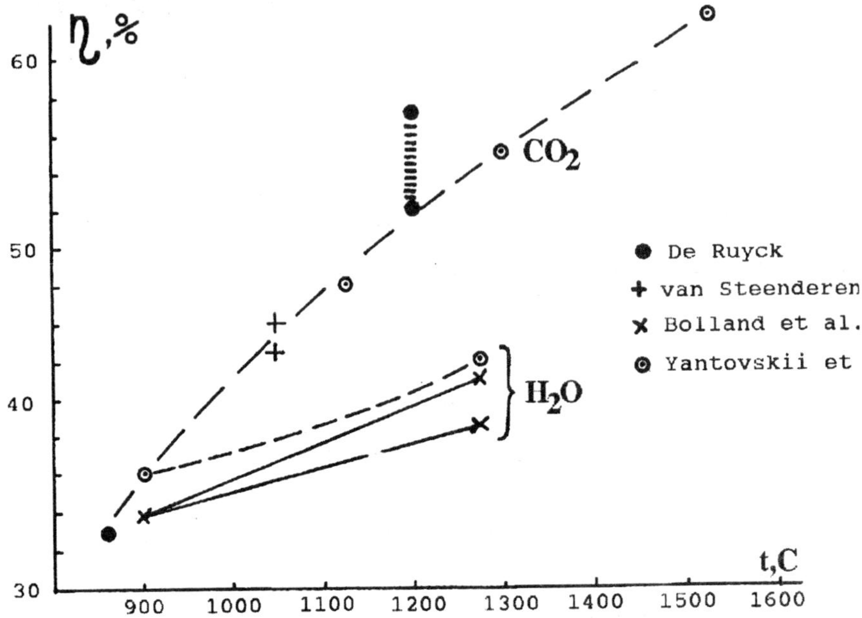

Figure 6.26 Comparison of estimated efficiencies versus temperature for all the mentioned schemes. In spite of different assumptions the general trend seems to be evident, see upper dotted line for carbon dioxide cycles.

Automotive Perspectives in Zero-Emission Cars

When speaking on the greenhouse effects we have to keep in mind that it is long range problem. Some controversies still exist whether it is a real problem or an invention of nuclear power proponents.

There exists a real problem indeed, it is urban and road air pollution. Let us citate:

James Cannon stated at the National Association of Fleet Administrators (NAFA) in Houston in April, 1990 that "air pollution in the United States is a national crisis. Nine people die every hour from exposure to air pollution." The American Lung Association recently confirmed earlier findings that, "most reasonable estimate for the total health cost from exposure to air pollution by the population of the United States lies in the range of tens of billions of dollars annually." As the debates of global warming continue, President Bush has proposed a 57% increase in the fiscal 1991 budget for climate change studies, increasing the total multi-agency funding to over 1 billion dollars for this effort.

Now in USA, only the annual toxic NO_X and hydrocarbons exhaust reaches 25 megatons respectively and forecast shows further increase.

The most active in air pollution is transportation (nearly 70%). Sometimes as the remedy the electric car with charged batteries is discussed. However, if it were widely used, the carbon dioxide would have increased due to additional power demand on coal- or gas-fired power plants. Only a solar-electric car could be considered as absolutely clean. But very low solar energy current density prevents direct use of this clean source.

The good accumulator of solar energy, hydrogen, certainly is much better than ordinary hydrocarbon fuels, however, the hydrogen combustion remains the toxic nitrogen oxides, therefore this can not solve the problem. It is evident from the table (Linch, Egan) of measured emissions of hydrogen fuelled car and bus:

gram per mile	HC	CO	NOx
Rosa Bus, 8890lbs	0,015	0,043	1,393
Rosa Auto, 2200lbs	0,004	0,011	0,345
Federal Standards	0,41	3,4	1,0
California Standards	0,41	7,0	0,4

Traces of HC and CO in hydrogen engine are produced from lubricating oil. The nitrogen oxide might exceed the emission standard.

Hitherto known alternative fuels (methanol, ethanol, reformulated hydrocarbons) give the same amount of toxics as ordinary gasoline.

The natural gas (methane) only seems to be better. Great firsthand experience got Italy already (Gambino et al.). There are 300,000 circulating vehicles with a methane fueling system. This number is destined to increase in the near future thanks to the interest of SNAM (Methane pipelines Administrators National Society) in the realization of new filling stations. The transformation of a car is carried out without any intervention on the engine, by installing a system of methane fuelling.

The test results obtained, Alfa Romero 1,712 liter (A.R. 1,7 engine) in grams per test are as follows:

	HC	CO	NO_x	CO_2	effic.
Gasoline	3,69/1,30	20,2/0,90	3,56/0,62	466/495	0,117
Methane	2,93/2,47	8,7/0,54	3,81/0,26	350/359	0,148

Aside to good influence of catalyst (second figure) we can see here some increase of toxic hydrocarbons in the methane case.

The main benefit of methane is its price is much less than gasoline. It is quite an abundant fuel indeed and its use seems to be widening.

Here exists the possibility to get a very clean vehicle if it is equipped by the engine or turbine using mentioned ICWEG cycle.

Certainly, the air separation device has to be installed on the vehicle, there exist wide variety of such devices to get oxygen. The replacement of nitrogen by carbon dioxide likely change nothing in working process of the engine. For the CO_2 compression, the part of cylinders might be used along with intermediate cooling.

Let us consider the Automotive Steam Turbine Without Exhaust Gases (AMSTWEG) cycle for a liquid gas fuelled car. Such a car contains a tank for a propane-butane gas mixture or even compressed methane. Certainly, in the case of methane, the mass of CO_2 effluent exceeds by up to 2.75 times that of the fuel; but the density of CO_2 is greater and, thus, the volumes of fuel and CO_2 are comparable. Hence, neutral liquid CO_2 separated by a soft membrane might fill the space of the consumed fuel.

A recuperator is not needed here since one large expansion in the turbine is used from 1000°C and 240 bars to 80°C and 0.5 bars (see Fig. 6.6). Table 6.4 shows the nondimensional breakdown of flow rates for the various constituents prior to this expansion and after combustion.

Table 6.4 Nondimensional flow rates for the AMSTWEG cycle.

	before combustion	after combustion
fuel	0.077	0.0
O_2	0.307	0.0
H_2O	0.616	0.789
CO_2	0.0	0.211
total	1.0	1.0

Note that the extra water in the case of this cycle could be used as make-up for the vehicle or elsewhere.

The cycle efficiency is given by

$$\frac{(4.56-2.64)(.875)-0.330-0.087}{4.56-0.33}=0.30$$

This is similar to the efficiency of modern Otto cycle engines. Of course, such engines have exhausts.

In the expansion work calculation, account was taken of the mass share of CO_2 which was 0.211 and which only contributes 0.086 to the total work of expansion due to the low gas constant. The correction factor in total work is (0.789+0.086)=0.875.

The above mentioned AMSTWEG cycle is based on steam as working substance. The same cycle on CO_2 might have some benefits due to efficiency increase.

The turbine supercharger of heavy-duty diesels is an example to transform it into compressor of CO_2. The liquid CO_2 with toxic contaminants dissolved in it as hydrocarbons, CO and NO_x is to be stored in the same fuel tank. It is discharged on filling station simultaneously when methane is refuelled.

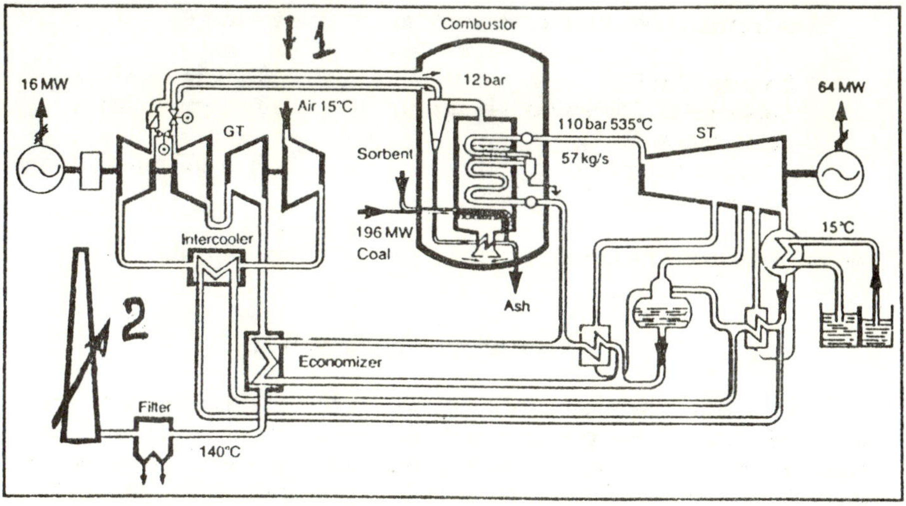

Figure 6.27 Flowsheet of the Pressurized Fluidized Bed power plant working in Stockholm, developed by ABB Carbon. The way to conversion for Pressurize Oxygen Carbon Dioxide Fluidized Bed Without Exhaust Gases (POCDFWEG): 1-install air splitting machine and mixer CO_2/oxygen before cycle compressor, 2-replace this chimney by the two pipes (one for extra CO_2 release in liquid form, another one for return CO_2 to the mixing with oxygen).

The density of natural gas, compressed up to 250 bars in the filling station is not more than 200 kg/cub.m (the methane critical density is 160 kg/cub.m). The density of liquid CO_2 by fuel temperature is not less than 550 kg/cub.m The mass relation CO_2/CH_4=44/16=2,75. It is evident that in any case the volume of methane fuel and stored CO_2 is the same, hence the tank might be the same.

From filling station, the liquid CO_2 is transferred by pipelines or lorries to dispose in deep wells to be dissolved in underground brines.

Certainly, methane-fuelled car is still too rare occasion to be the object of major efforts in a zero-emission problem. We have to find a solution for arbitrary fuels.

Here the OXICE project might be of interest. This engine works on arbitrary hydrocarbon fuel and a mixture oxygen-carbon dioxide instead of air is used. After combustion and cooling in radiator, the small amount of clean water is easily separated and dry carbon dioxide is compressed and stored in the fuel tank.

The key problem is oxygen if we can't carry air splitting plant on board due to its large mass. The oxygen filled capsule with solid oxygen (oxygen ice=OXICE) seems to be better solution. Solid oxygen is quite explosion proof; it can only sublime without explosion, in contrast to liquid one.

The preliminary data of this car are in the table:

Power 70 HP	Tank mass 55 kg
Mileage 12,5 Miles/kg (44 mpg)	Total mass 135 kg
Fuel charge 20 kg	
Oxygen charge 80 kg, 55 1	Run 250 miles/charge
Carbon dioxide 55 kg, 47 1	

For refuelling one has to discharge CO_2 and change capsule.

Pressurized-Fluidized Bed

The use of coal for ICWEG leads to the gasification and Pressurized Fluidized Bed. These plants are especially promising in use of low grade fuels, where large amount of ashes has to be disposed of. In some particular cases, like Estonian shales, the ash is strongly alkaline and large lifeless ponds and lakes appear near the power plants. It is the good place to dissolve here CO_2 to neutralize alkalines getting carbonates and to return life in these basins.

The PFB technology is well developed already in Stockholm Vartan power plant with two units of 100 MW each. It is the experience of great importance to the world power production. The plant is clean indeed, with its emission of nitrogen and sulfur oxides on the level of 10 ppm. However the CO_2 emission exists there. The further development might be as it is shown on the Fig. 6.27 to install air splitting machine for the $02/CO_2$ mixture and to liquify extra CO_2 for disposal of. The pressures and temperature in the PFB reactor might be increased (if the turbines would withstand), because NO_x is dissolved in CO_2 completely.

Exergonomics of the OCDOPUS Project

(Oil enhancement Carbon Dioxide Oxygen Power Universal Supply)

The particular aim of the plant is carbon dioxide supply for enhancement of oil recovery along with compressed nitrogen, if necessary.

Present paragraph is aimed at trying to evaluate effectiveness of mentioned OCDOPUS plant by means of exergy flow calculations in all the "arms." The use of exergy is unavoidable as in many multipurpose plants having released flows of different kind. Exergy efficiency η_ε can be calculated easily as a relation of delivered exergy flow to its income.

In a simplified manner all the flows from and to are presented on the Figure 6.28. Here just eight arms are evident. Certainly, some cooling water is needed too for a cooling. In case of a drilling platform it is sea water.

Note, however, that the plant cannot produce all flows simultaneously. For example the compressed nitrogen production (7 kg/s, 300 bar) needs to consume all 10 MW of power.

The plant has some modes of operation: for power, for nitrogen or mixed. Liquid CO_2, process steam and hot water are inherent effluents of the unit.

The plant outline is presented on the Figure 6.29. The things of our project are restricted within dotted line boundary.

Air enters the separation unit 1, where it is splitted on the two flows, oxygen and nitrogen. It is depicted as a small square, but in effect a separation unit is a big facility, its mass is as much as for the whole power plant. The oxygen enters combustion chamber 2 along with recirculated CO_2 and flow of fuel gases. Combustion products (mainly CO_2) expand in the turbine 8, cooled down in the recuperator 3 and cooling tower 4. After the water separation in 5 the dry CO_2 is compressed from 7 up to 60 bar in a compressor 10, with additional recuperators.

The CO_2 condensation at 60 bar takes place in 11, then liquid CO2 is pumped up to 100 bar and returned to recuperator 3.

Combustion born water might be heated in 9 if a process steam is needed. Combustion born CO_2 is released in liquid state and enters a pipeline to be injected in a well.

The nitrogen compressor 6 is attached to the shaft by a juncture, it might be switched on or off on site. When the total flowrate of nitrogen from separation unit is to be compressed and injected the net power to the grid equals zero.

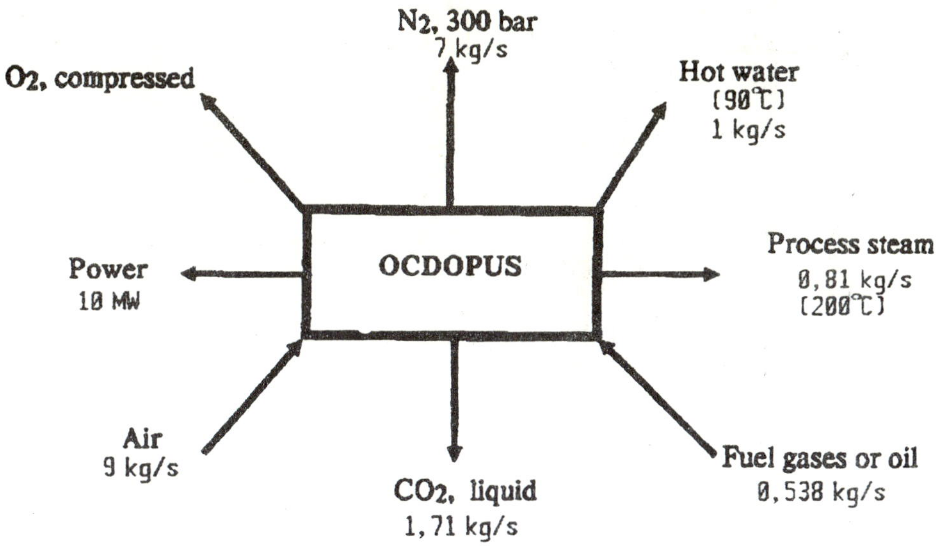

Figure 6.28 OCDOPUS production and consumption.

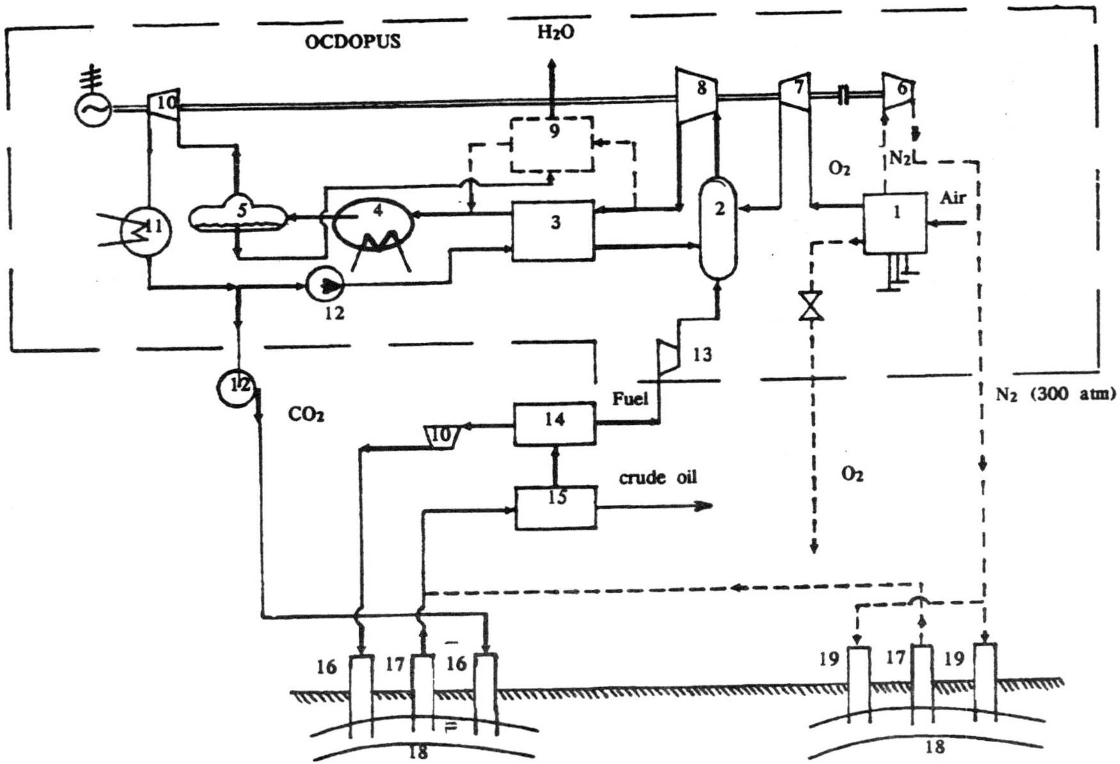

Figure 6.29 1-Unit for air separation; 2-Combustion chamber; 3-Recuperator; 4-Cooling tower; 5-Water separator; 6-Nitrogen compressor; 7-Oxygen compressor; 8-Turbine; 9-Steam production; 10-CO_2 compressor; 11-CO_2 condensor; 12-CO_2 pump; 13-Fuel gas compressor; 14-CO_2/fuel gas separator; 15-Oil/gases separator; 16-CO_2 injection well; 17-Productive well; 18-Reservoir; 19-Nitrogen injection.

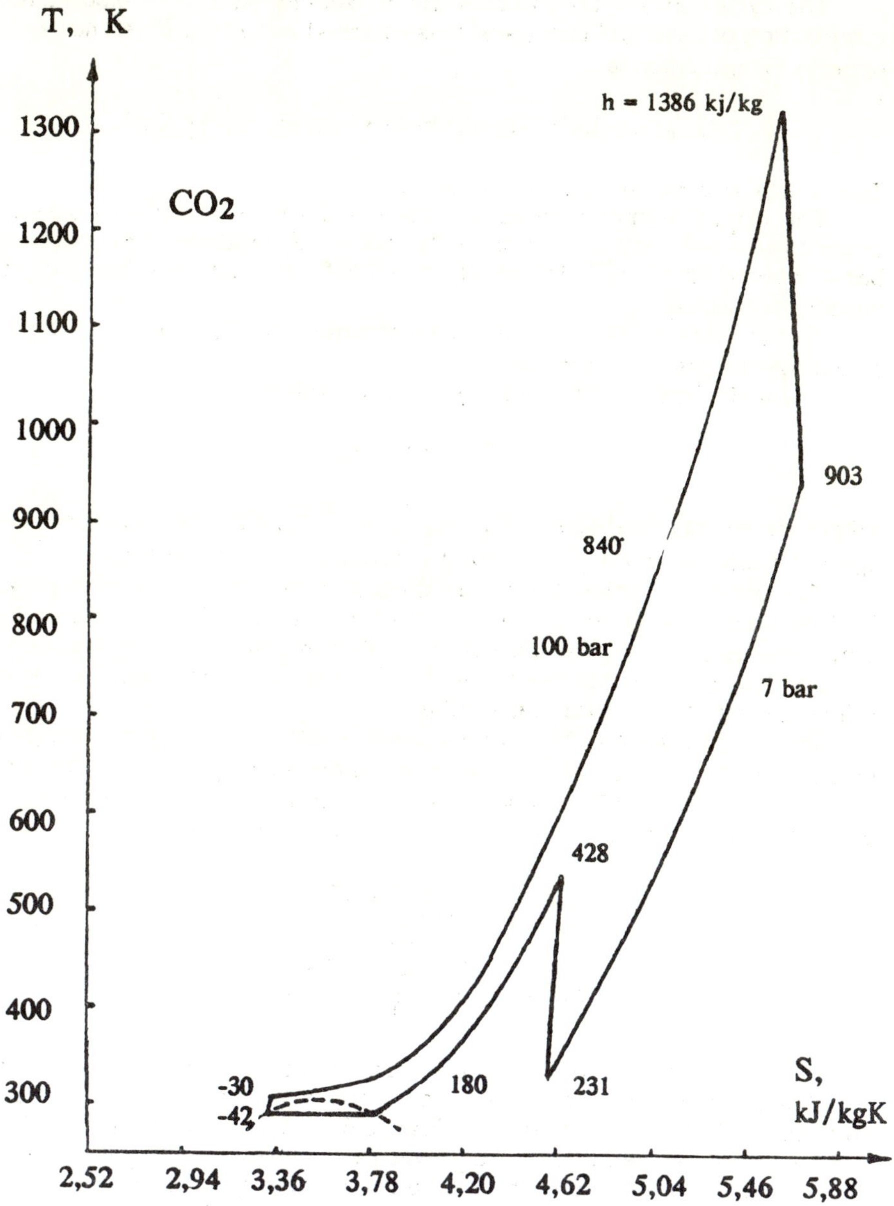

Figure 6.30 T-S diagram of the cycle with adiabatic compression and additional recuperators.

The cycles in T-S coordinates are presented on Figure 6.30. Chemical composition of used fuel (oil gases) was assumed as C_6H_{14}. Reference equation of perfect combustion is:

$$1/6\ C_6H_{14}+19/12\ O_2 \rightarrow CO_2+7/6\ H_2O+50\ MJ/kg\ C_6H_{14}$$

The stoicheometric coefficient equals 3.6.

The upper temperature of the cycle 1323 K (1050°C) was selected in order to use ordinary gas turbines "of-the-shelf." Rather high pressure 100 bar is needed in CO_2 cycles to explore the thermodynamic benefits of this working substance.

Turbine isoentropic efficiency was assumed as 0.85, it is a real figure for a cooling-free gas turbine blade.

The main criterion of exergonomics is as follows:

$$Z=1/\eta_\varepsilon+1/K_\varepsilon$$

where net-exergy coefficient $K_\varepsilon = \dot{\varepsilon}_{del}\tau_n / \varepsilon_{inv}$; $\dot{\varepsilon}_{del}$-flow of delivered exergy; τ_n-life time (or normative); ε_{inv}-total amount of invested exergy.

The ordinary Sankey-Grassman diagrams for exergy flows are presented in Figure 6.31. Exergy income is fuel. Exergy of air considered as zero. Exergy outflow in electrical power measured by the same amount of watts. Exergy of compressed nitrogen has been calculated by a simple formula for ideal gas. Minor exergy of hot water was omitted.

For CO_2 the data of Szargut are used. It was verified by summing up the exergy of CO_2 from dry air by 1 bar and compression up to 60 bar, by 293 K.

The two modes of operation, for power and for compressed nitrogen are presented on diagram. Related efficiencies are as follows:

$$\eta_\varepsilon^{power} = \frac{10+1.48+0.75}{29} = 0.42$$

$$\eta_\varepsilon^{N_2} = \frac{3.47+1.48+0.75}{29} = 0.20$$

The main problem of net-exergy calculations consists of mass evaluations of all the components of not existing facility. We make use of quite a big statistics collected by Zaritski, see Figure 6.32. Here specific mass of heavy-duty gas turbines versus specific power per flowrate unit is presented. There exists correlation: the more specific power the less specific mass.

In OCDOPUS project the total flowrate equals 53.8 kg/s and specific power is 10,000/53.8=185 kJ/kg. Having looked at Figure 6.32 one can see our appraisal for power plant near to upper figures: 10 kg/kW, it means 100 t for 10 MW power.

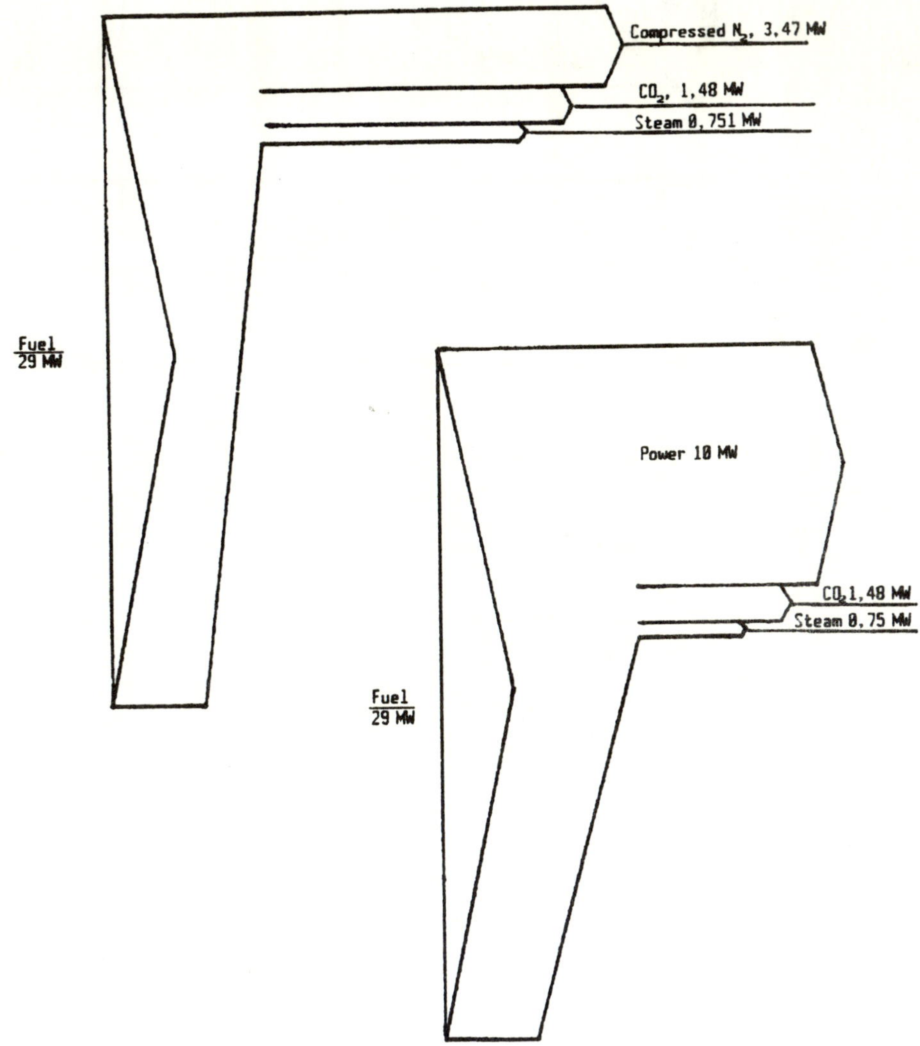

Figure 6.31 Exergy flow diagrams of OCDOPUS.

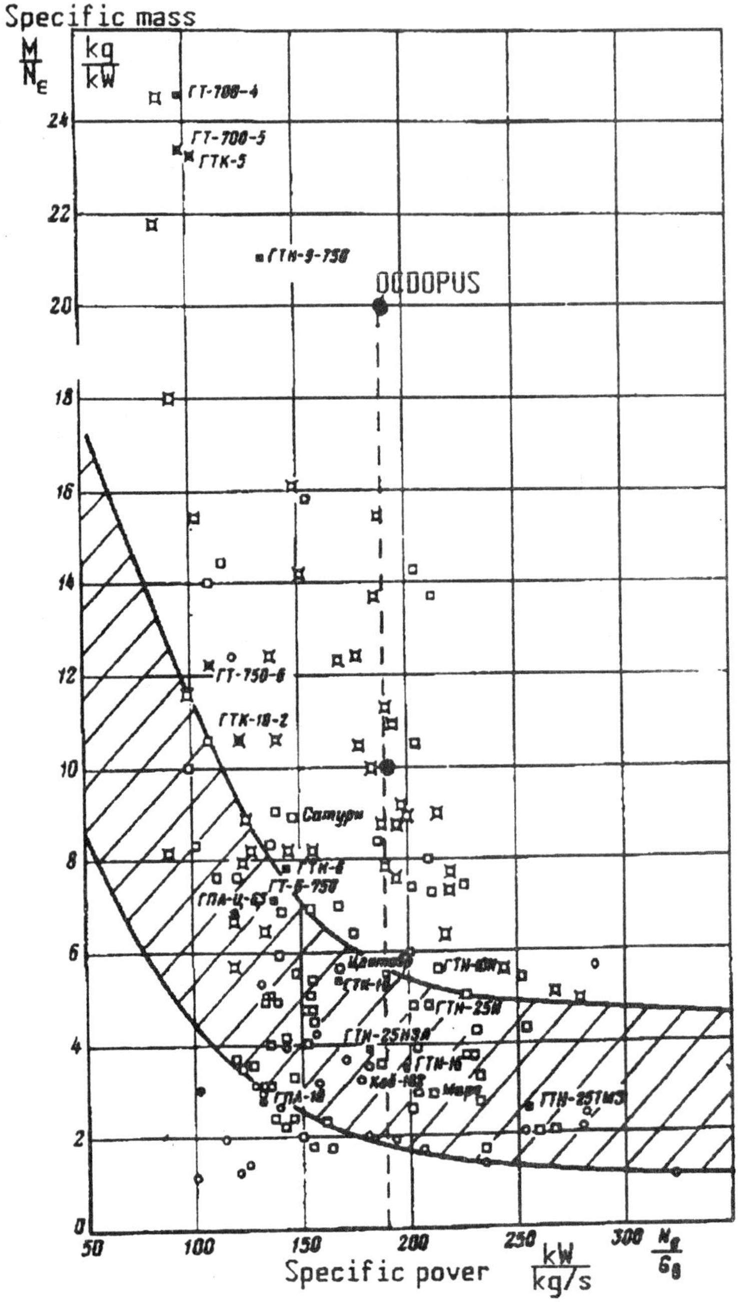

Figure 6.32 Heavy-duty gas turbine statistics and OCDOPUS point.

Averaged data for oxygen plants are around specific mass of 50 t/kg O_2, it gives the same mass for air separation unit 100 t because oxygen flowrate equals 2 g/s. Therefore total mass of OCDOPUS is 200 t or 20 kg/kW.

Exergy intensity of steel, heavily stainless, is about 100 MJ/kg. Therefore total exergy investment is around 200.100=20 TJ.

Delivered exergy per normative time τ_n=8 years and 20 megasec of working time in a year equals:

$$10^6 \cdot 8.2 \cdot 10^7 = 1600 \text{ TJ}$$

Net-exergy coefficient K_ε is equal 1600/20=80. It is very high value of K_ε therefore the role of invested exergy in Z criterion is small as in many fuel-fired power plants. If it is approved by more rigorous calculations the optimization of OCDOPUS for maximal η_ε only will be justified.

All the calculations for OCDOPUS project were done by my students J. Tryggstad and L. Lindquist (Sweden).

Chapter 7.

EXERGONOMICS AND GREENHOUSE EFFECT

A prudent man forseeth the evil, and hideth himself; but the simple pass on, and are punished.

Proverbs, 27:12

In the previous chapter some new thermodynamic cycles and outlines of power plants have been discussed.

In chapter 5 we have seen one option to prevent global disaster related to desertification of the planet surface.

There exists, however, another menace, global warming and glacier melting, leading to increase of the level of World Oceans. Sometimes the guesses appear that the warming might enhance the desert attack. The aim of this chapter is to describe the physics of possible warming due to "greenhouse effect" and then to offer remedy.

Physics of Greenhouse Effect

All the gases absorb electromagnetic radiation of very short waves. Different gases used to do this on different wave bands, fixed to any particular gas (see Fig. 7.1) The shortest wave of ultraviolet an visible light from the Sun reaches the surface of our planet with minor moderation. The very high temperature of the Sun leads to radiation of major part of energy in this interval of wave lengths. The surface of the Earth (much less heated than the Sun) can irradiate only long waves of infrared radiation into space, which might be absorbed by atmosphere due to presence of some actively absorbing gases. The peak of reradiated energy from the Earth occupies the length of some tens of micrometers. It is just the band of absorbing capacity of carbon dioxide. This lead to return of energy current back to planet surface, preventing its own radiation. The CO_2 acts like a blanket around the planet. Other gases like steam absorb waves far from the peak of reradiated energy. They present much less harm to global warming. However, there exists the sum of other gases like methane, chorfluorcarbons and other, which acts as a whole like CO_2, being responsible for the remained half of the effect.

The following table presents our collection of data, mostly of Marchuk et al., which shows the effect of reflected from the atmosphere radiation.

Surface Temperatures and Energy Currents, W/sq.m.

Planets	$\frac{t_{av}}{t_{min}+t_{max}}$	Solar Constant	Surface	Long w.	Reflected
Moon	$\frac{-18}{-150,+100}$	1360	$\frac{1360}{15+6000}$	$\frac{1360}{15+6000}$	0
Earth	+15°C	1360	390	237	150
Venus	+460°C	2620	17900	130	17770

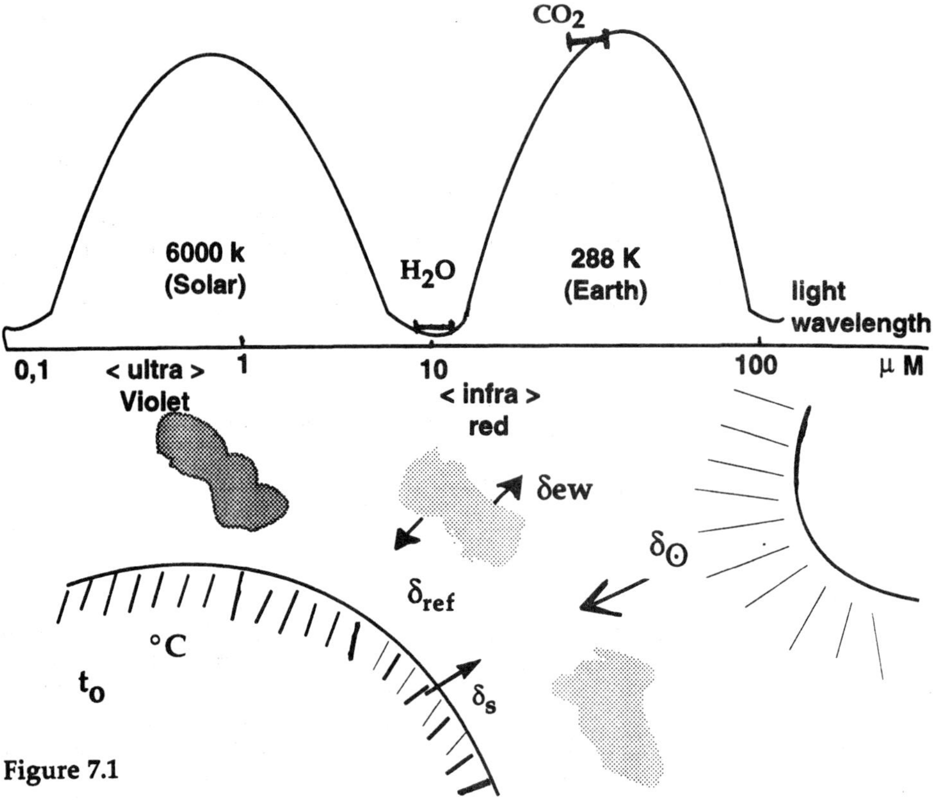

Figure 7.1

Here are commensurated three well known planets, two of them with atmosphere (Earth, Venus) and one without (Moon). The long wave radiation is going into space from outer layers of atmosphere. The reflected radiation 150 W/sq.m, going back to Earth is responsible for the elevation of average temperature (+15°C) over that on the Moon (-18°C). This radiation might be increased if the content of CO_2 and other trace gases would increase.

The last line related to Venus vividly shows what may happen if the process will go too far. The surface temperature of Venus forbids any life (460°C). Its origin is CO_2 rich atmosphere and very high energy current of reflected radiation.

Venus is the ultimate result of greenhouse warming. Is the Earth going to Venus setting? Many things discovered recently give a positive answer.

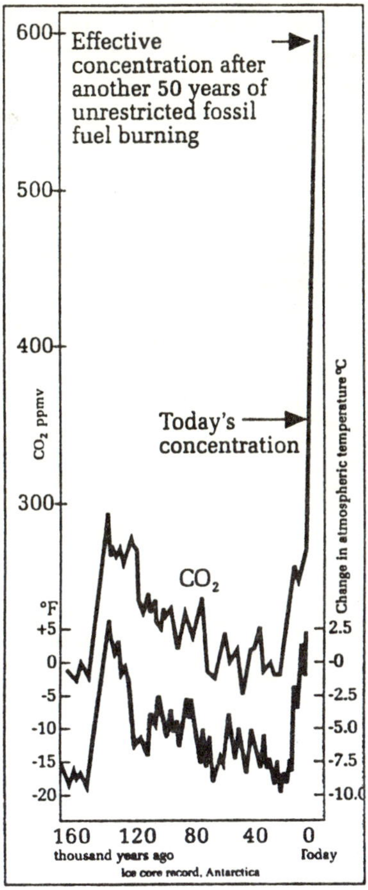

Figure 7.2

This graph shows the close correlation between two lines. The upper one reflects the content of CO_2, measured by spectral, very sensitive methods in the bubbles of air inside a thick layer of ice in Antarctica. A very long time span has been covered; 160 thousand years from present time.

The lower one presents the change in atmospheric temperature along this time span,which has been registered in that ice layer (Greenpeace leaflet).

What is to be done?

In many recent books, especially in the Krause, Bach and Koomey monograph, a lot of data confirming the reality of warming have been presented. The very impressive is rather old (1981) picture of Hansen, where gradual improvement of models has been shown (Fig. 7.3). The more sophisticated model, the better agreement between the calculated curve and observations. The growing consensus confirms warming menace and urgent necessity to lessen man-made CO_2 release into atmosphere.

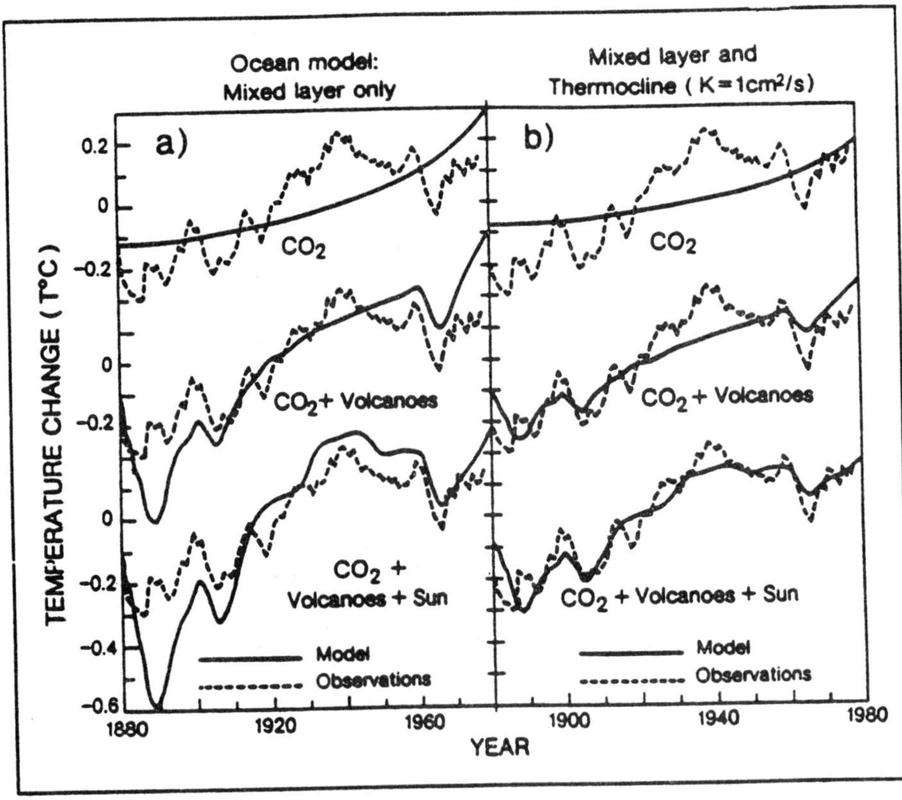

Figure 7.3 Global temperature trend obtained from a global climate model for several assumed radiative forcings.

At first glance, the man-made CO_2 production is minor in comparison to natural sources, recalculated per carbon it means 5 Gt annually versus 100 Gt/y exchange atmosphere-ocean, see Fig. 7.4 (Sarlos and Verstraete). But detailed calculations show that this 5% are in effect crucial.

What is to be done? CO_2 and other greenhouse gases abatement. The global target is presented in Fig. 7.5 (Krause). Many fold decrease of the release rate is needed, for low-level risk only 15% should remain. Is it possible? We will not discuss the switch to nuclear power. It is the same as to prevent headache by means of cancer without headache.

There might be three ways for further development: fuel conservation, fuel substitution by renewables and stack gases sequestering.

To conserve fuel, to lessen combustion gases formation is only one way: to increase exergy efficiency.

Here, however, a principal question should be answered, which partially relates to economics (or econometrics).

Ordinary economic discounting is based on a decreasing weight function $\exp(-\lambda t)$, where the time t is dimensionless due to the relation to the total life time of a considered object and λ might be an inversed time span, sometimes referred to as a normative time.

As we have seen in Chapter 4, the quality of a project is estimated by the sum of specific exergy consumption Z for the current supply and for the construction of an equipment, which is similar to the ordinary cost of energy.

Here appears the crucial question, which can not be answered within the limits of thermodynamics:

Should we discount exergy like money?

If yes, what is here the λ parameter or the time $1/\lambda$?

The social will, which is considered by the present writer as the subject of econometrics instead of physics, should be taken into account.

In contemporary evaluations of the scale M of fuel conservation a simple formula is widely used:

$$M=(1-\eta)B \tag{7.1}$$

where η=exergy efficiency, B=total fuel consumption. There exists, however (see Chapter 4) an optimal η. When we try to get η more than optimal one, in effect energy is wasted due to the increase of an exergy investment to create equipment. More realistic formula:

$$M=(\eta_{opt}-\eta)B \tag{7.2}$$

Here the first term depends strongly on the parameter λ, its variation from 0.25 to 1 leads to changing of the optimal η from 0.4 to 0.8 in the most typical cases, when η=0.2. The change of a possible fuel conservation here is threefold.

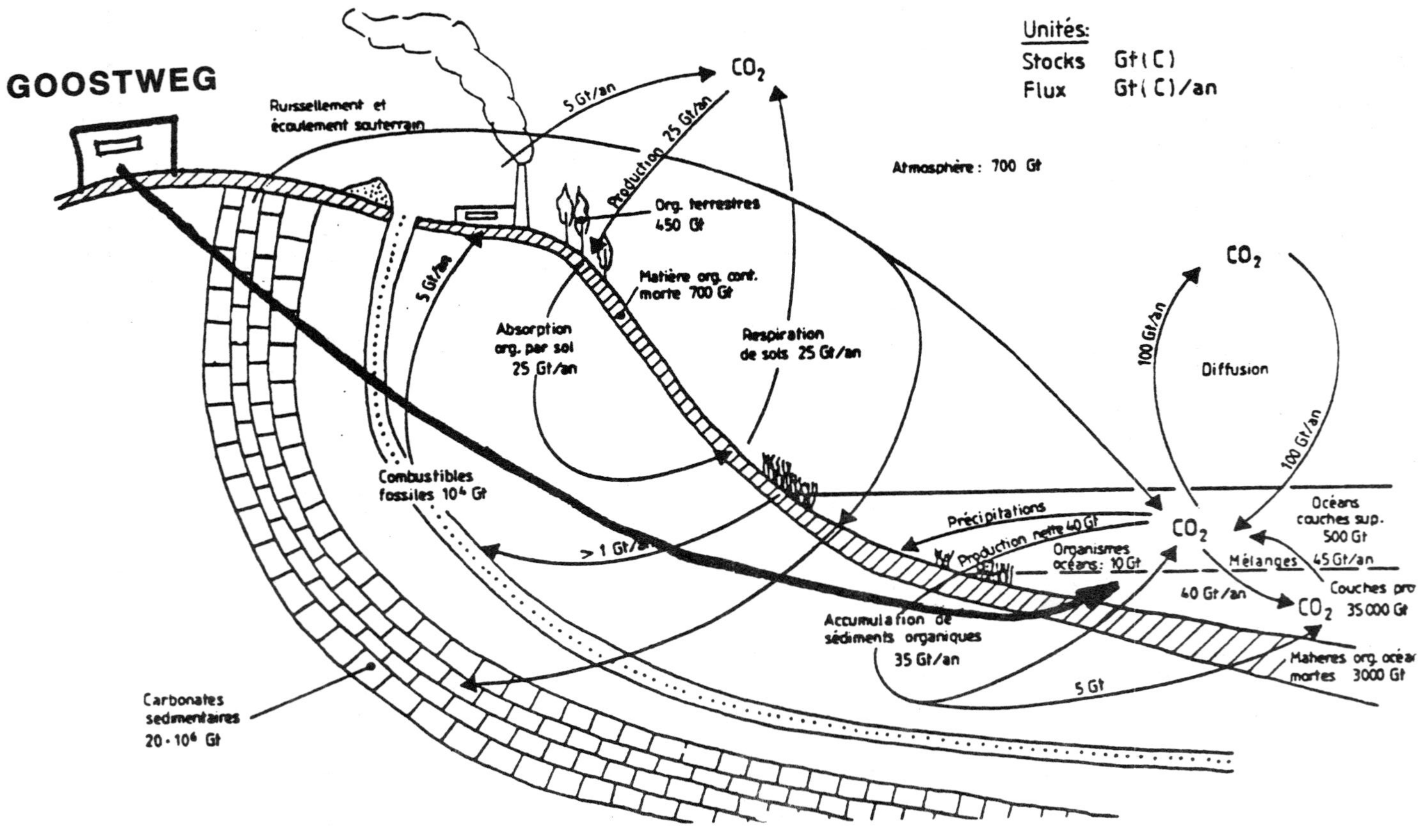
GOOSTWEG
Ruissellement et écoulement souterrain
5 Gt/an
CO_2
Production 25 Gt/an
Unités:
Stocks Gt(C)
Flux Gt(C)/an
Atmosphère : 700 Gt
Org. terrestres 450 Gt
Matière org. cont. morte 700 Gt
5 Gt/an
Absorption org. par sol 25 Gt/an
Respiration de sols 25 Gt/an
CO_2
100 Gt/an
Diffusion
100 Gt/an
Combustibles fossiles 10^4 Gt
> 1 Gt/an
Précipitations
Production nette 40 Gt
Organismes océans: 10 Gt
CO_2
Océans couches sup. 500 Gt
Mélanges
45 Gt/an
40 Gt/an
Couches pro
CO_2 35 000 Gt
Accumulation de sédiments organiques 35 Gt/an
5 Gt
Matières org. océan mortes 3000 Gt
Carbonates sédimentaires $20 \cdot 10^6$ Gt

Figure 7.4

TABLE 2 : Quantities of carbon in the Earth's crust and the atmosphere

Carbone en circulation	
CO_2 ou CO_3 en solution	Gt
Atmosphère	700
Oceans	
— couches supérieures	500
— couches profondes	35 000
Carbone organique	
Organismes	
— terrestres	460
— marins	10
Matières organiques décomposées	
— terrestres	700
— marines	3 000
Total C en circulation	environ 40 000 Gt
Carbone fixe	
Carbonates sédimentaires	20.10^6
Combustibles fossiles	10 000
Total C fixe	environ 20.10^6 Gt

Figure 7.4 *(continued)*

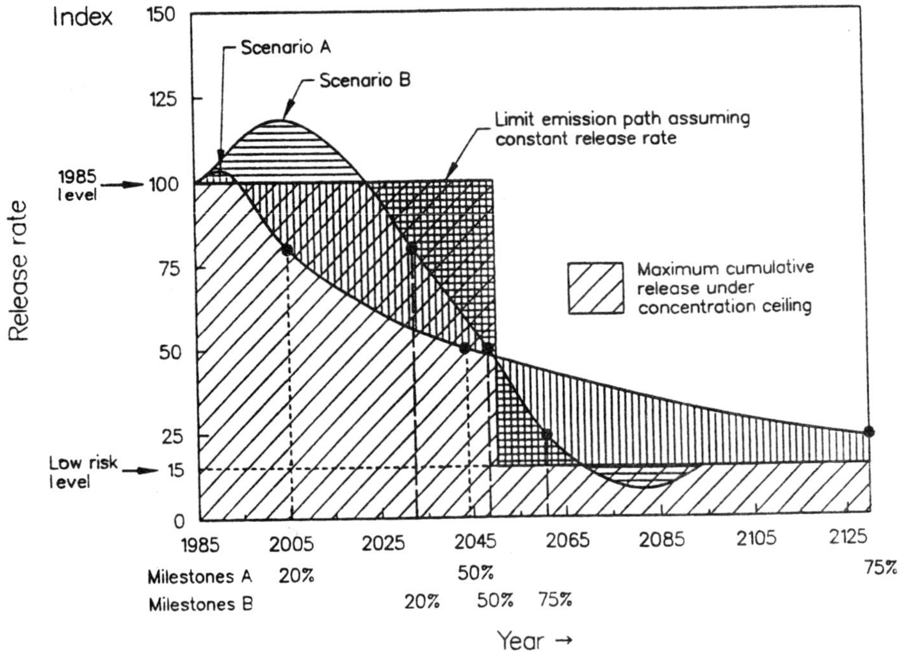

Figure 7.5 Climate stabilization targets: relationship of concentration ceilings, emission paths, and reduction milestones.

The very important conclusion on the impossibility to meet the Toronto target in the carbon dioxide abatement appears if $\lambda=0.25$. Otherwise, if the $\lambda=1$, it is probably possible.

A less known but more general formula (from Chapter 4) is as follows:

$$M=(Z_{real}-Z_{min})B/Z_{real} \tag{7.3}$$

Here we need to know the λ as well, because Z_{min} is the function of λ.

If econometrics were able to help exergonomics to choose the parameter λ, it would be an advance in both sciences.

The question is still open. In a realistic case we cannot hope for a possibility to meet the target of Fig. 7.5 by means of fuel conservation due to exergy efficiency increase.

An attentive reader may object me: should we use the efficiency η_{tot} in the formula (7.1) everything would be correct, the exergy investment there are taken into account. The thing is, however, that η_{tot} versus η_{ε} has a maximum, it never can approach to unity, therefore the different $(1-\eta_{tot})$ is much higher than may expect a proponent of the formula (7.1). In this approach the (7.3) is valid as well.

The fuel substitution by renewable, clean energy sources is certainly the right way. In the long-range perspective, the solar-hydrogen option seems to be the best (Bockris, 1980). We do not see any physical obstacle to develop it. However, it might not come sufficiently soon. Many decades are needed to widen the use of renewables.

But CO_2 abatement is needed urgently in near decades. Here remain the only way, CO_2 sequestering. The deep ocean layers and underground oceans are suitable for stack gases disposal. The principle solution presented on the Fig. 7.4 using acronym "GOOSTWEG," has been considered in some detail in the Chapter 6.

Country	Contribution to world CO_2 emissions	Policy plans
Australia		20% reduction by 2005
Belgium	1.6	Stabilize at the 1988 levels by 2000
Canada	0.5	Stabilize at the 1990 levels by 2000
Denmark	2.0	20% reduction by 2005
Finland	0.3	Stabilize at the 1990 levels by 2000
France	0.3	Recommend 20% reduction by 2005
Germany	1.9	30% reduct. on the 1987 levels by
Ireland	3.2	2005
Italy	0.1	Stabilize at the current level by
Japan	1.8	2000
Netherlands	4.4	Stabilize at the 1990 levels by 2000
New Zealand	0.7	Stabilize at the 1990 levels by 2000
Norway	0.1	Stabilize by 1995
Sweden	0.2	20% cut by 2000
Switzerland	0.2	Stabilize at the 1990 levels by 2000
UK	0.2	Stabilize at the 1990 levels by 2000
USA	2.8	20% cut proposed
	22.0	Stabilize at the 1990 levels by 2005
USSR		Currently no policy plans
	18.4	
		Currently no policy plans

Figure 7.6

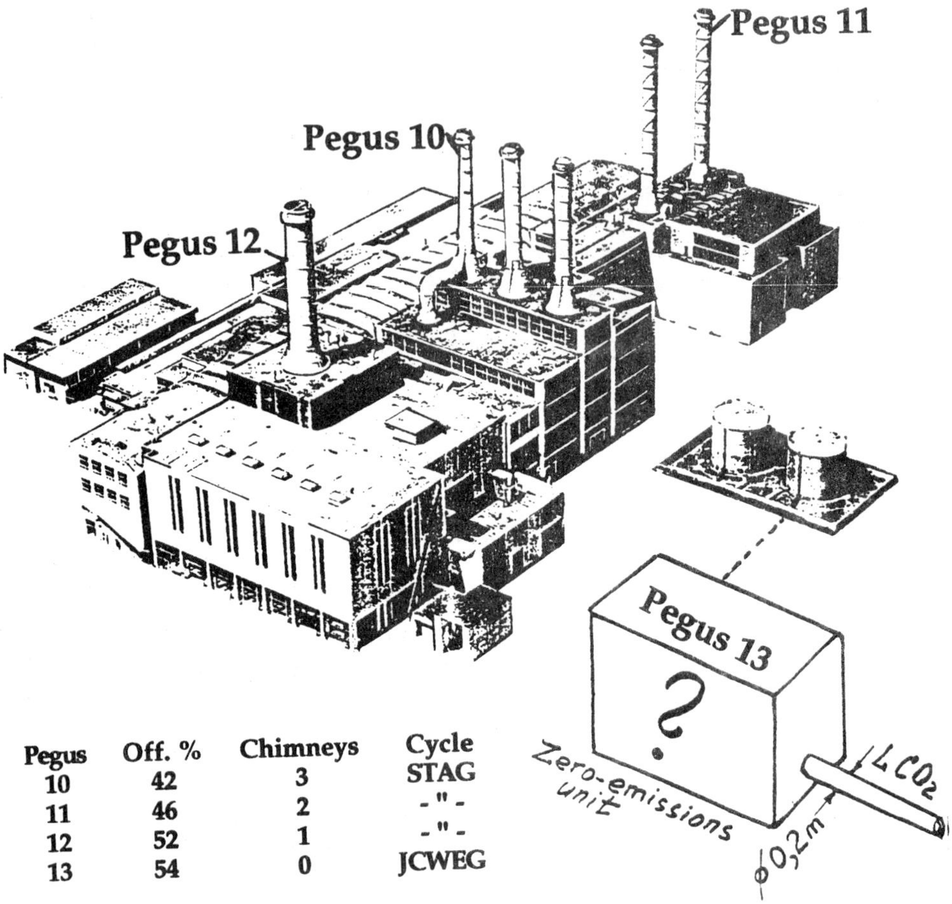

Pegus	Off. %	Chimneys	Cycle
10	42	3	STAG
11	46	2	- " -
12	52	1	- " -
13	54	0	JCWEG

Figure 7.7 **Merwedehanaal Unit 12, Utrecht**

The most efficient way seems to be the retrofit of fuel-fired power plants in the direction of Internal Combustion Without Exhaust Gases (JCWEG), which has been presented in the preceeding chapter. Where is the best point to start the job?

Let us consider the distribution of CO_2 emission over the list of countries (Fig. 7.6). There is evidence that greatest role of big countries USA and the late USSR, in harmful emission. The share of other individual countries is minor. However, here, the great importance of an example appears.

In such small, very well developed country as the Netherlands, the total stop of emissions gives minor effect on a global level. But the Dutch people have made their country by themselves, without even the Lord God's assistance, they can start. The purpose for this start seems to be one of the best power plants in a world, Merwedekanaal plant of UNA utility in the city or Utrecht. At the picture (Fig. 7.7) of this book the outlook of this plant is presented. The recently installed units Pegus 10 (3 chimneys), Pegus 11 (2 chimneys), Pegus 12 (1 chimney) show gradual increase in efficiency up to 52%. The natural question emerges how many chimneys would be in the Pegus 13? The evident answer is: zero.

Here the writer might be asked: you have considered many clean power cycles in Chapter 6, you promised to recommend the remedy. Could you select the best approach by yourself?

In reply I wish to present the detailed results of computer simulations made with K.N. Zvagolsky and V.A. Gavrilenko. Making it we always kept in mind the Pegus units in Utrecht Merwedekanaal power plant, trying to give in an expicit manner our offer for the Pegus-13. It is referred to as the COOPERATE cycle.

The fuel of the cycle is natural gas (methane) as is the case for all the Pegus units. The cycle seems to be really clean, zero-emissions, it is of high thermodynamic level due to carnotization. It means of minimized energy losses in all the parts of the cycle.

The COOPERATE cycle: (CO_2 Prevented Emissions Recuperative Advanced Turbines Energy)

In the plant scheme Figure 7.8 there are only minor changes in respect to previous chapters. Air enters the separating unit 1, the produced oxygen after compression from 1 to 240 bar enters the first combustion chamber 2 along with fuel and recirculated heated carbon dioxide. After triple combustion and expansion the flue gases are cooling down in recuperator 3 and in a cooling tower.

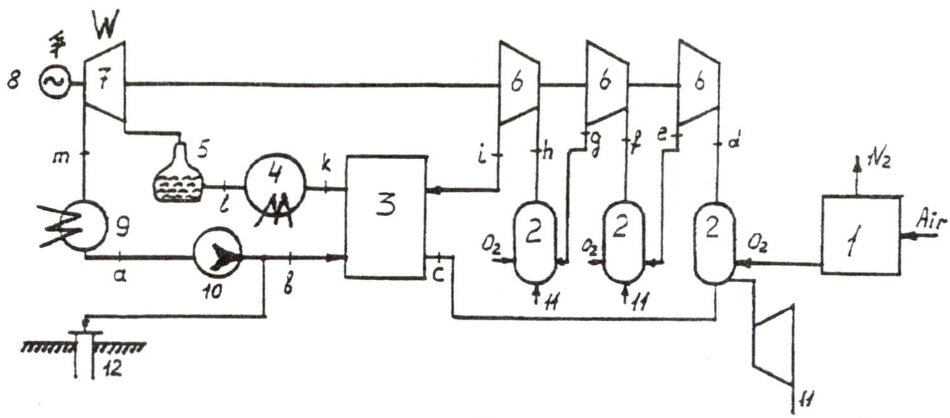

Figure 7.8 Outline of COOPERATE cycle: 1- air separation unit; 2- combustion chamber; 3- recuperator; 4- cooling tower; 5- water separator; 6- turbine; 7- intercooled compressor; 8- generator; 9- CO_2 condensor; 10- CO_2 pump; 11- fuel; 12- depleted well.

After the water separation the almost dry CO_2 enters the multistaged intercooled compressor 7 and condensor 8, where by pressure 60 bar and 20 C the liquefaction takes place. Liquid CO_2 is pumped from 60 to 240 bar and enters recuperator 3.

In view of possible pollution of CO_2 by some contaminants the liquid CO_2 is treated by chemicals for the same reason as a water treatment device is used at the steam ordinary power plant.

For the mixture of combustion products CO_2+H_2O (flue gases) playing the role of working substance we need to know the enthalpy H(t,p) and entropy S(t,p) which are additive functions of that of pure CO_2 and H_2O:

$$H=H_{2}O^{*}X_{CO_2}+H_{H_2O}{}^{*}X_{H_2O}$$
$$S=S_{CO_2}{}^{*}X_{CO_2}+S_{H_2O}{}^{*}X_{H_2O}$$
$$X_{CO_2}+X_{H_2O}=1$$

The function H(t,p) and S(t,p) assumed to be polynoms, the coefficients have been calculated by nonlinear regression based on the least square method. The mean inaccuracy of the data equals 0.003-0.8% in the range of p=300 bar, t=20-1500 C.

Numerical example for one point t=1300 C and p=15 bar is as follows. Mass fraction of water is 0.053, of CO_2 is 0.947. The pure CO_2 and water from (Altunin, Rivkin) have:

H=2309 and 5410 kJ/kg
S=6.228 and 9.538 kJ/kg*K

The properties of mixture are:

H=2309*0.947+5410*0.053=2473 kJ/kg
S=6.228*0.947+9.358*0.053=6.394 kJ/kg*K.

The difference to pure CO_2 properties is evident.

Reference equation of equilibrium combustion reaction and molecular masses are as follows:

$$CH_4+2O_2=CO_2+2H_2O+50\ MJ/kg\ CH_4$$
16 64 44 36

The primary information of the results is presented on the Tables 7.1-7.8. The first four of them are for the turbine inlet temperatures of 950, 1000, 1250 and 1300 C, pressures 240/60/15/4 bar, see Tables 7.1-7.4. Note that temperatures t_e, t_g and t_i are indicated for isoentropic expansion by the presented values of entropy. The enthalpies in the same points are calculated for appropriate entropy rise according to efficiencies 0.8, 0.85 and 0.9. The iteration process was rather quick, from 5 to 23 iterations.

Table 7.1

mCH_4 0.025	mO_2 0.100	mCO_2 0.069	mH_2O 0.056	$(1-mCO_2-mH_2O)$ 0.874
	P bar	T °C	H kJ/kg	S kJ/(kg*K)
a	60	20	566	
b	240	40	586	
c	240	519	1296	
d	240	950	1926	5.3366
e	60	658	1627	
f	60	950	1967	5.7197
g	15	721	1713	
h	15	950	1999	6.0130
i	4	628	1620	
k	4	50	930	
l	4	30		

Iterations=5
EFFICIENCY=0.469

Hde=281 L_{CH_4}=17
Hfg=247 L_{CO_2}=17
Hhi=378 Hcd=630
D_{O_2}=72 Hef=340
D_{CO_2}=166 Hgh=285
L_{O_2}=44

Table 7.2

mCH_4 0.026	mO_2 0.103	mCO_2 0.071	mH_2O 0.058	$(1-mCO_2-mH_2O)$ 0.871
	P bar	T °C	H kJ/kg	S kJ/(kg*K)
a	60	20	566	
b	240	40	586	
c	240	565	1356	
d	240	1000	1995	5.3940
e	60	702	1686	
f	60	1000	2037	5.7751
g	15	765	1773	
h	15	1000	2071	6.0699
i	4	668	1676	
k	4	50	931	
l	4	30		

Iterations=5
EFFICIENCY=0.481

Hde=289 L_{CH_4}=17
Hfg=256 L_{CO_2}=17
Hhi=395 Hcd=639
D_{O_2}=74 Hef=351
D_{CO_2}=166 Hgh=298
L_{O_2}=45

Table 7.3

mCH_4 0.030	mO_2 0.118	mCO_2 0.081	mH_2O 0.066	$(1\text{-}mCO_2\text{-}mH_2O)$ 0.852
	P bar	T °C	H kJ/kg	S kJ/(kg*K)
a	60	20	566	
b	240	40	586	
c	240	798	1654	
d	240	1250	2343	5.6543
e	60	914	1981	
f	60	1250	2397	6.0245
g	15	973	2074	
h	15	1250	2444	6.3265
i	4	860	1957	
k	4	50	945	
l	4	30		

Iterations=15 EFFICIENCY=0.539

Hde=334 L_{CH_4}=19
Hfg=311 L_{CO_2}=17
Hhi=487 Hcd=689
D_{O_2}=85 Hef=416
D_{CO_2}=165 Hgh=370
L_{O_2}=51

Table 7.4

mCH_4 0.030	mO_2 0.121	mCO_2 0.083	mH_2O 0.068	$(1\text{-}mCO_2\text{-}mH_2O)$ 0.849
	P bar	T °C	H kJ/kg	S kJ/(kg*K)
a	60	20	566	
b	240	40	586	
c	240	843	1712	
d	240	1300	2412	5.7047
e	60	959	2043	
f	60	1300	2470	6.0700
g	15	1014	2134	
h	15	1300	2520	6.3723
i	4	896	2013	
k	4	50	951	
l	4	30		

Iterations=23 EFFICIENCY=0.548

Hde=339 L_{CH_4}=20
Hfg=323 L_{CO_2}=17
Hhi=508 Hcd=700
D_{O_2}=87 Hef=427
D_{CO_2}=164 Hgh=386
L_{O_2}=52

After selection of turbine inlet temperatures all the other quantities are calculated by the ordinary mass and energy conservation equations. The mass fractions of fuel, oxygen, steam and recirculated CO_2 are calculated simultaneously.

The t-S diagram for the case of t_d=t_i=t_h=1250 C is presented on Figure 7.9. The lower part of the cycle is almost independent of mentioned temperatures of upper part provided the pressures are fixed.

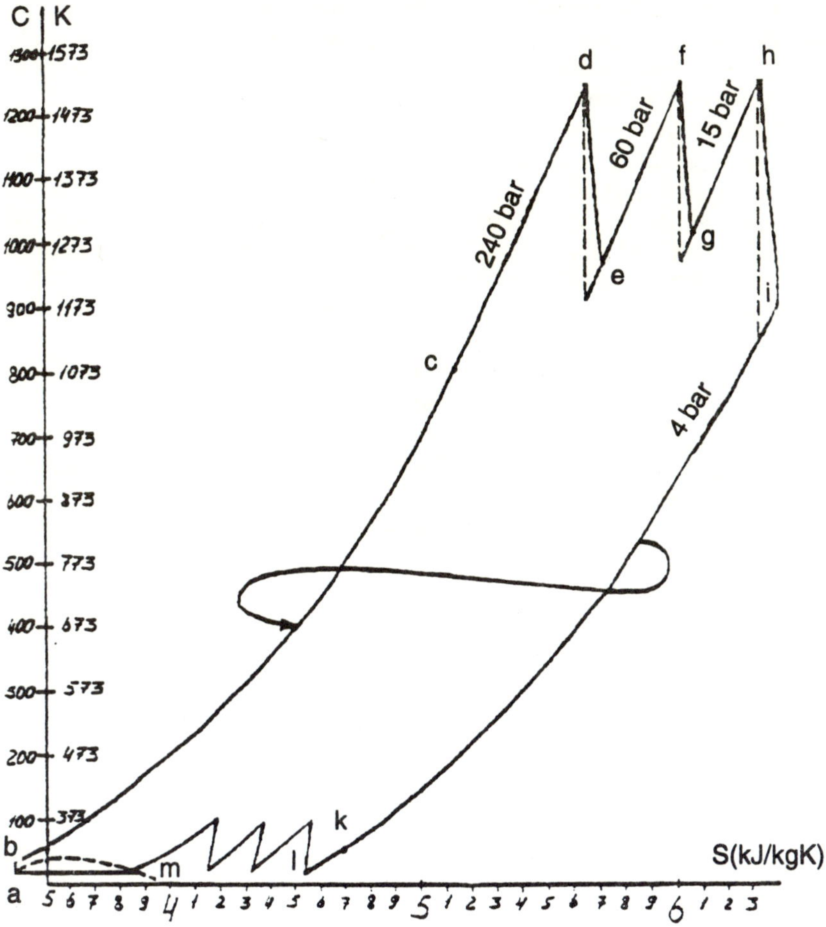

Figure 7.9 The COOPERATE cycle in t-S coordinates

The cycle efficiency is calculated as follows:

$$\eta = \frac{L_1 + L_2 + L_3 - (D_{O_2} + D_{CO_2} + L_{O_2} + L_{CH_4} + L_{CO_2}}{H_{cd} + H_{ef} + H_{gh}} =$$

$$\frac{334 + 311 + 487 - (85 + 165 + 51 + 19 + 17)}{689 + 416 + 370} = 0.539.$$

Air separation power consumption assumed as 0.2 kW/kg O_2, oxygen compression up to 240 bar was isothermal by compressor efficiency of 0.8.

Carbon dioxide compression was calculated as the sum of adiabatics by the efficiency 0.8, with pressure ratio of 2 in every stage.

Specific power of CO_2 pumping is $L_{CQ}=(p_b-p_a)/\rho\eta_L$ by efficiency $\eta_L=0.6$.

The calculated cycle efficiency, turbine power and fuel mass fractions versus turbine temperatures are presented on Figure 7.10. Even by lowest temperatures efficiency is rather high due to effective recuperation (0.469). At the highest end it reaches 0.552 by 1300 C. The trend toward saturation of the curve is evident.

The ratio η/η_c where $\eta_c=1-T_0/T_d$, T0=293 K, reflects effectiveness of the coversion of thermal energy flow (heat) into power. It is just exergy efficiency of heat transformation but not the total exergy efficiency which coincides to η due to small difference of methane exergy in respect to heating value. The ratio of η/η_c is very high due to small temperature drops in recuperator (from 50 to 10 C). Ratio is increased with the temperature rise due to decrease (t_i-t_c) which is a major source of exergy losses.

The graphs on Figure 7.10 give some information to a main decision in a power plant design, the choice of the upper cycle temperature. The higher it is, the higher is efficiency, however the problems of turbine blade materials, cooling of blades and recuperator construction are increased too.

The distribution of temperatures over the turbines might be uneven. We can lessen the temperature of the first, very high pressure turbine to lessen engineering problems while maintaining the last low pressure turbine at high temperature. The example of uneven distribution (1100, 1200 and 1300 C) is presented in Table 7.5. In respect to even 1200 C distribution efficiency remained the same (0.527 and 0.528), however in uneven case recuperator temperature t_c=834 C instead of 752 C. It involves some difficultues in recuperator design. At first glance such uneven distribution is not promising, however it needs more calculations and seems to be an interesting subject of future investigations.

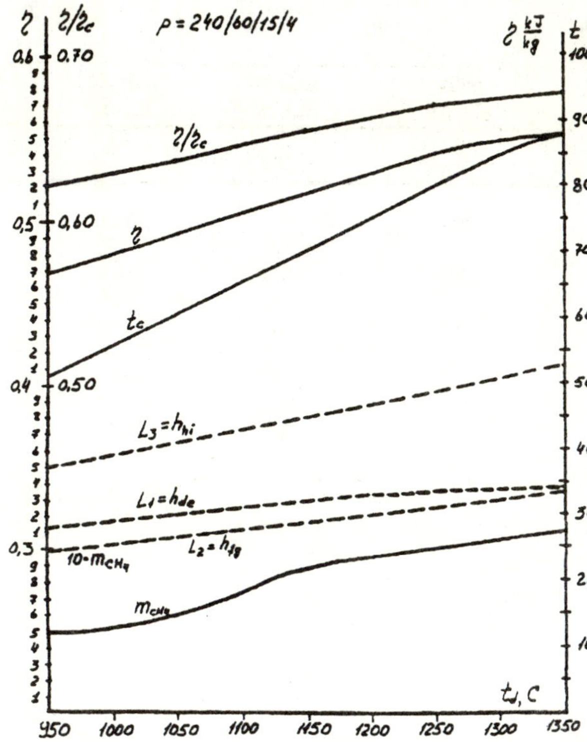

Figure 7.10 The cycle efficiency η, recuperation temperature tc, specific turbines power L and fuel mass fraction m versus turbines inlet temperature.

An important problem is the back pressure of the last turbine selection. The major part of present calculations relates to p_i=4. The cycle efficiency is sufficiently high by the small dimensions of recuperator due to high heat transfer coefficient in a dense flue gases.

The thing, however is the possibility to increase cycle efficiency due to deeper expansion in the last turbine. Tables 7.6-7.8 present the calculated values versus turbine temperature by the back pressure bar. The latter case gives higher efficiency up to 0.588 and lower t_c=716 by the highest t_d=1300 C. It means that decrease of back pressure gives not only efficiency rise but also the recuperator problems ease.

Table 7.5

mCH_4 0.029	mO_2 0.115	mCO_2 0.079	mH_2O 0.065	$(1-mCO_2-mH_2O)$ 0.856
	P bar	T °C	H kJ/kg	S kJ/(kg*K)
a	60	20	566	
b	240	40	586	
c	240	834	1701	
d	240	1100	2101	5.4805
e.	60	790	1777	
f	60	1200	2304	5.9610
g	15	932	1996	
h	15	1300	2511	6.3627
i	4	896	2005	
k	4	50	944	
l	4	30		

Iterations=n.a EFFICIENCY=0.527

Hde=291 L_{CH_4}=17
Hfg=292 L_{CO_2}=17
Hhi=506 Hcd=400
D_{O_2}=83 Hef=528
D_{CO_2}=165 Hgh=514
L_{O_2}=47

Table 7.6

mCH_4 0.031	mO_2 0.123	mCO_2 0.085	mH_2O 0.069	$(1-mCO_2-mH_2O)$ 0.846
	P bar	T °C	H kJ/kg	S kJ/(kg*K)
a	60	20	566	
b	240	40	586	
c	240	543	1328	
d	240	1100	2158	5.5179
e	60	785	1822	
f	60	1100	2202	5.8999
g	15	849	1913	
h	15	1100	2241	6.2016
i	1	611	1652	
k	1	50	954	
l	1	30		

Iterations=7 EFFICIENCY=0.534

Hde=313 L_{CH_4}=21
Hfg=279 L_{CO_2}=17
Hhi=589 Hcd=830
D_{O_2}=89 Hef=380
D_{CO_2}=179 Hgh=327
L_{O_2}=55

Table 7.7

mCH_4 0.033	mO_2 0.130	mCO_2 0.089	mH_2O 0.073	$(1-mCO_2-mH_2O)$ 0.837
	P bar	T °C	H kJ/kg	S kJ/(kg*K)
a	60	20	566	
b	240	40	586	
c	240	630	1440	
d	240	1200	2300	5.6216
e	60	868	1941	
f	60	1200	2349	6.0002
g	15	932	2036	
h	15	1200	2393	6.3045
i	1	679	1753	
k	1	50	959	
l	1	30		

Iterations=9 EFFICIENCY=0.557

Hde=332 L_{CH_4}=22
Hfg=302 L_{CO_2}=17
Hhi=640 Hcd=861
D_{O_2}=94 Hef=408
D_{CO_2}=178 Hgh=357
L_{O_2}=58

Table 7.8

mCH_4 0.034	mO_2 0.137	mCO_2 0.094	mH_2O 0.077	$(1-mCO_2-mH_2O)$ 0.829
	P bar	T °C	H kJ/kg	S kJ/(kg*K)
a	60	20	566	
b	240	40	586	
c	240	716	1550	
d	240	1300	2440	5.7248
e	60	958	2067	
f	60	1300	2498	6.0945
g	15	1013	2160	
h	15	1300	2548	6.3989
i	1	744	1853	
k	1	50	965	
l	1	30		

Iterations=12 EFFICIENCY=0.578

Hde=342 L_{CH_4}=23
Hfg=326 L_{CO_2}=17
Hhi=695 Hcd=890
D_{O_2}=98 Hef=431
D_{CO_2}=177 Hgh=388
L_{O_2}=61

The natural gas-fired power plant of 1 GW by the inlet temperature 1250 C and efficiency 0.539 consumes 37 kg/s of fuel plus 148 kg/s of oxygen. The flow-rate of released liquids is 100 kg/s of CO_2 and 85 kg of fresh, combustion born water.

The best known gas turbine cycles like I-STIG using GE turbine LM 8000 of design output 114 MW by inlet temperature 1360 C does have efficiency 48%, consuming 27 kG/s of injected steam which is exhausted along with other stack gases. For 1 GW power these water losses are 237 kg/s. One can see that COOPERATE cycle promises higher efficiency and production of fresh water instead of consumption in I-STIG cycle.

Now we need to answer the crucial question "when". When could we request or select the turbines, combustors, recuperators and other equipment?

As to air separation we can be certain; technology is mature. Bug manufacturing companies like Air Products & Chemicals, Inc. can immediately deliver all the air separation systems (see booklet "Cryogenic Air Separation Systems" 1988, printed in USA 220-701).

Here we have to distinguish the two kinds of technologies:

	Cryogenic	Chemical (MOLTOX System)
Economic Range	>20t/day	>500t/day
Purity limit	99.8%	99.9%

The main advantage of the MOLTOX system (using an oxygen-absorbing mixture of alkali salts) is a 30-40% efficiency gain over a caparable cryogenic plant.

The process is best suited for customers with a high volume, high purity oxygen requirement above 500 t/day and a need for excess steam. It fits the need of COOPERATE cycle, its efficiency might increase by two points due to a decline of oxygen production power.

As to turbines and combustors, the only problem is just the first turbine and combustor. There are no developed turbines having steam-like pressures (240 bar) and gas turbine temperatures around 1300C simultaneously.

For immediate realization we offer the COOPERATE demo version of the cycle, see Figure 7.11. Here we use the flexibility of the triple-expansion cycle with uneven temperature distribution over the turbines. The first combustion chamber is eliminated and the first expansion takes place just after the recuperator. Pressure after the first turbine falls down to 60 or 30 bar. These gases are heated either in an additional recuperator or in an ordinary gas-turbine combustion chamber, then driving the ordinary or slight modified gas turbines.

Having looked at the T-S diagram of Figure 7.12 one can consider the cycle as a combined one, however here steam-like turbine forms not bottoming but topping part of the cycle. The power of it very much exceeds the pumping power of liquid CO_2, giving a good thermodynamic income.

For the described COOPERATE-demo version all the equipment seems to be developed already. So the question "when" might be answered: very soon, upon request.

In respect to Pegus-12 unit the COOPERATE cycle promises even higher efficiency by the same temperatures before turbine, however its benefit is evident - cleanliness. Therefore the mentioned remedy is COOPERATE power plant.

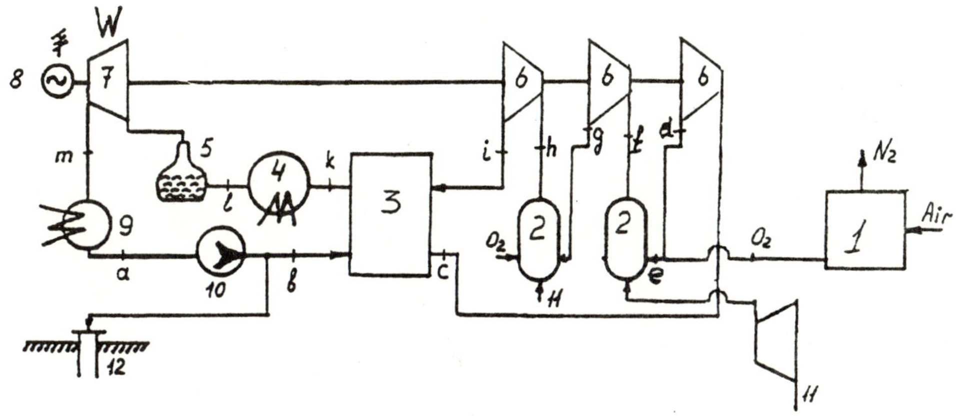

Figure 7.11 Flowsheet of COOPERATE-demo version. 1-Air separation unit; 2-Combustion chamber; 3-Recuperator; 4-Cooling Tower; 5-Water seperator; 6-Turbine; 7-Intercooled compressor; 8- Generator; 9-CO_2 condenser; 10-CO_2 pump; 11- Fuel; 12-Depleted well.

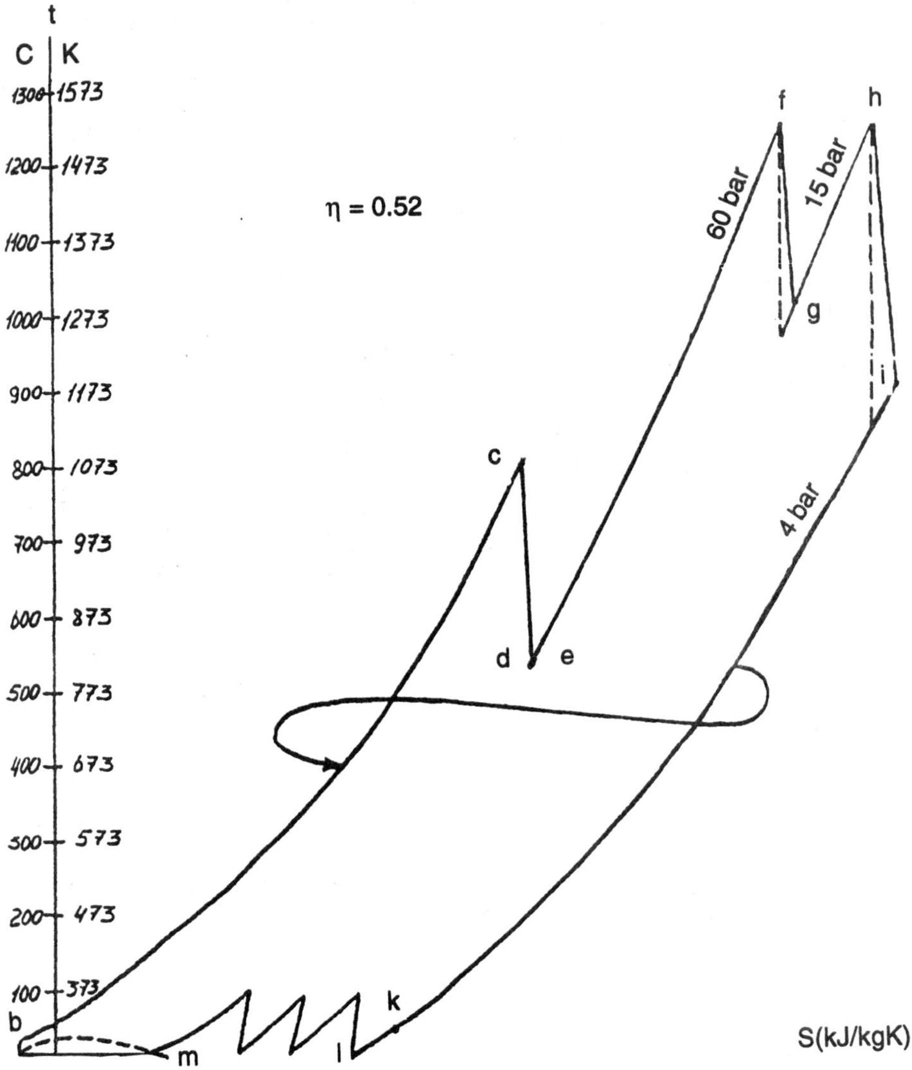

Figure 7.12 The COOPERATE-demo cycle on T-S diagram.

Conclusion

When we look at shortages of resources, we often say that not only energy is restricted, but also ore in the ground, clean water, food and so on.

In reply it can be pointed out that due to mass conservation law, no one chemical element could escape from the Earth. What results indeed is as follows: water became dirty, metals are oxidized, hydrocarbons are burning. From a thermodynamic point of view all these processes are, in effect, examples of entropy rise and exergy loss.

If the firm exergy source is found, like solar energy, we can always reduce iron of oxides, to clean water and to synthesize hydrocarbons.

It is evident that exergy is just that absolute resource which has to be conserved.

Energy production and conservation is in effect the same as exergy when speaking in everyday life. Energy and exergy currents always coexist. They might be equal as in the case of electrical or mechanical power, as well as strongly differing in the heat processes. The flow of exergy might be greater than that of energy; it may be directed inversely with respect to energy flow. Therefore it is wrong to speak of exergy as a part of energy. It is connected to energy but quite another function.

The exergy flow diagrams, with explicit image of transfer of exergy from one carrier to another, are a useful tool in energy processes design. Exergy efficiency is clearly presented as a relation of the outflow to inflow.

The exergonomics, as an attempt to generalize exergy analysis and match it to information and risk analysis, gives a graphical image of the figurative point trajectory and the base of multicriterial (Pareto) optimization.

The main criterion of exergonomics, sum of specific exergy consumption, may play a role of a target function in the optimization calculations, especially for future projects when economic prices are unknown.

Some energy and exergy flow case studies in energy projects show their benefits as well as bottlenecks and sore points.

The preliminary calculations of energy and exergy flows in solar ponds, electrical heaters, clean· power cycles of JCWEG type, and coal-fired in oxygen MHD generator, show their promising income to the solution of clean energy problem.

In view of urgent need to protect the atmosphere of our planet, the internal combustion cycles, without exhaust gases, should be investigated.

The construction of the cycles presented here requires air splitting devices with capacities of up to 100 kg/s of oxygen for power plants and 10 g/s for car engines. A decrease in power consumption is desirable but not necessary. In fact, in spite of the significant power consumption for air splitting and CO_2 liquefaction, all the cycles presented here have efficiencies similar to those of their exiting counterparts. The primary benefit of these cycles is, however, evident - CLEANLINESS. The disposal of liquid carbon dioxide and the contaminants dissolved in it can be done without harm to the environment.

It would be good to remember here the conclusion edited by W. Chandler of a joint comprehensive investigation. It is as follows:

> *What can the world's leaders do to reduce the risk of climatic change? A global climate convention will be necessary to set enforceable goals and provide specific mechanisms for achieving them. A goal of at least holding emissions constant could be pursued with little economic impact. The priorities for specific action include:*
>
> *Setting efficiency standards at cost-effective levels for appliances and automobiles.*
>
> *Shifting taxes to energy resource use.*
>
> *Shifting energy research funds to develop greater end-use efficiency.*
>
> *Shifting energy research to the long-term development of noncarbon fuels.*
>
> *Sharing technology for efficiency with Eastern Europe, the Soviet Union, and developing countries through joint research developing financial institutions, and promoting joint ventures.*

Of course, all of these words are by all means correct but are as insufficient as a prayer. *And all things whatsoever ye shall ask in prayer, ye shall receive* (Matthew 21:22). In order to protect the atmosphere, the immediate engineering action must be to develop energy supplies without emissions to the atmosphere. The cycles presented here might be a step in the right direction.

The way is long and hard. Energy and exergy current evaluations play the role of lighthouses on that way.

References

Alessio F.J., Energy Analysis and the Energy Theory of Value, Energy Journal, 1982, v.2, No1., c.61-74.

Alexejev, V.V., Liamin, M. Ya., Shirokova, E.L., *Biomass of Microalgae use for Solar Energy Conversion,* Techno-economic and Ecology Aspects of Ocean Energy Use, TOI Vladivostok, pp. 53-58, in Russian, 1985.

Altfeld K., Leiner W., Fiebig, M. Second Law Optimization of Flat Plate Solar Air Heaters, Solar Energy Engineering. 1988, Part 1, v.41, No2, p. 127-132.

Altfeld K., Leiner W., Fiebig, M. Second Law Optimization of Flat Plate Solar Air Heaters, Solar Energy Engineering, 1988, Part 2, v.41, No4, p. 309-317.

Altunin V.V. Thermophysical Properties of Carbon Dioxide. Moscow, Standard Publishers, 1975 (in Russian).

ATHENS '91 = Analysis of Thermal and Energy Systems, Proc. of the International Conference, Athens, Greece, June 3-6, 1991, Greg Foundas Editions.

Babaev, N., Varfolomeev, A., Kuz'man, J., Hydrogen Energy Progress VII, v.3, p. 1779-1799, 1988. Proc. 7th World Hydrogen En. Conf.

Baehr, H.D., Die Exergie von Kohle und Heizol, BWK, 1987, v.39, No.1-2, p. 42-45.

Bailey, R.A., Net Energy Analysis of Eight Technologies to Provide Domestic Hot Water Heat, Energy, 1981, v.6, No.10, p. 983-997.

Bejan, A., Advanced Engineering Thermodynamics, NY, J. Wiley, 1988.

Benemann, J.R. Utilization of Carbon Dioxide from fossil fuel-fired power plants with biological systems. Energy Conv. and Mgmt., V. 34, No. 9-11, pp. 999-1004, 1993.

Blok, K., J. Farla, C.A. Hendris, W.C. Turkenburg, (1991). Carbon Dioxide Removal: A Review. ESETT'91, Milan, Oct. 21-25

Blok, K., Turkenburg, W., Hendriks, C., Steinberg, M. (ed.) (1992) Proceedings of the First Intern. Conf. on Carbon Dioxide Removal. Pergamon Press.

Bockris, J.O.M., Wass, J.C., About Real Economics of Massive Hydrogen Production, Proc. 7 World Hydrogen Energy Conf., v.1, Pergamon, 1988.

Bockris, J.O.M., Energy options, Tailor and Francis, 1980.

Bockris, J., *guest comment*, Environment Conservation, v. 17, No3, Autumn 1990.

Bolland, O., Saether S. New concepts on natural gas fired power plants which simplify the recovery of carbon dioxide. Energy Conversion & Mgmt., V. 33, No. 5-8, 1992, p. 467.

Boustead, J., Hancock, B., Handbook on the Industrial Energy Analysis, London, Ellis Horwood, 1979.

Brodiansky, V., Dolinsky, A. (ed), Exergy Calculation in Technical Systems. Handbook, Kiev, Naukova dumka, 1991 (in Russian).

Brown, L.M., Zeiler, K.G. Aquatic Biomass and Carbon Dioxide Trapping. Energy Convers. Mgmt. V. 34, No.9-11, pp. 1005-1013, 1993.

Chandler, W.U., *Carbon Emissions Control Strategies,* Case Studies in International Cooperation, World Wildlife Fund, Baltimore, Maryland, U.S.A., 1990.

Cleveland, C.J., Constanza, R., Hall, C.A.S., Kaufman, R., Energy and US Economy: A Biophysical Perspective, Science, 1984, v.225, p. 890-897.

Cleveland, C.J., Hereneen, R., Solar Parabolic Collectors: Successive Generation Are Better Net Energy and Exergy Producers, Energy Systems and Policy, 1989, v.13, p. 63-77.

Critopf, R.E., Energy Analysis, In: Energy-Present and Future Options, v.1, Ed. Merrick, Wyley and Sons, New York, 1982, p. 313-330.

deGroot, S.R., Termodynamica dei Processi Irreversibili, Societa Italiana di Fisica, Bologna (1960).

de Vries, B., Nieuwlaar, E., 1981, A Dynamic Cost-exergy Evaluation of Steam and Power Generation. Resources and Energy, v.3, p. 359-388.

Degtiarev, V.L., Grybovsky, V.P. Carbon dioxide semi-closed power plant. Author Sert USSR, No. 295 897 of July 28, 1967, published in Bullet. of Inventions Nov. 12, 1971, No. 8, F Olk 13/00, C01b 3/00.

DeRuyck, J. Efficient CO_2 capture through a combined steam and CO_2 gas turbine cycle. p. 397.

Ebeling, W., Volkenstein, M.V., "Entropy and Evolution of Biological Information," *Physics* A 163 (1990), pp. 398-402.

Ebeling, W., Engel-Herbert, H., Herzel, H., "Thermodynamic Aspects of Self-Organization," in *Self-Organization by Nonlinear Irreversible Processes*, Spriner, 1986, pp. 2-15.

ECOS '92 = Efficiency, Cost, Optimization and Simulation of Energy Systems, Proc. of the Intern. Symposium, Zaragoza, Spain, June 15-18, 1992, ASME.

ENSEC '93 = Energy Systems and Ecology, Proc. of the Intern. Conference, Cracow, Poland, July 5-9, 1993, Technical University of Silesia.

Evans, D.B., Energy Analysis as an Aid to Public Direction Making, Int. Chem. Eng. Symposium. Series No78, p. 23-36.

Frantz, C.C., Cambel, A.B., Net Energy Analysis of Space Power Satellites, Energy. 1981, v.6, p. 485-501.

Fratzcher, W., Brodjanskij, V.M., Michalek, K., Exergie, Theorie und Anwendung, Leipzig, VEB, Deutscher Verlag fur Grundstoffindustrie, 1986, p. 348.

Gambino, M., Iannaccone, S., Unich, A., Light Duty and Heavy Duty Spark Ignition Engines Fueled with Methane. Analysis of Thermal and Energy Systems. Proc. of the Intern. Conf., Athens, Greece, June 3-6, 1991, pp. 631-641.

Garcusha, L.K., Schegolev, G.M., *Entropy Diagrams for Combustion Products*, Naukova Dumka, Kiev, USSR, 1968.

Giovanninni, B., Pain, D., *Scientific and Technical Arguments for the Optimal Use of Energy* Proceedings of the 2nd Annual CMDC Conference, Zürich, Switzerland, December 10-11, 1990.

Gilliland, M.W., Klopatek, J.M., Hildebrand, S.G. Net Energy: Results for Small-scale Hydroelectric Power and Summary of Existing Analysis, Energy, 1981, v.6, No10, p. 1029-1040.

Haefele, W. (ed), *Energy in a Finite World*, N.Y. Ballinger, 1981.

Hannon, B., Thermodynamics: Principle and Applications. Chapter 5. Energy Discounting. In: Energetics and Systems. Mitsch W., et al. 1982, p. 73-97.

Hendriks, C.A., K. Blok, W.C. Turkenburg (1991) Technology and Cost of Recovering and Storing Carbon Dioxide From an Integrated Gasifier Combined-cycle Plant Energy, v.16, No11/12., pp. 1277-1293.

Herendeen, R.A., Plant, R.L., Energy Analysis of Four Geothermal Technologies, Energy, 1981. v.6, p. 73-82.

Herzog, H., Golomb, D., Zemba, S., Feasibility, Modeling and Economics of Sequestering Power Plant CO_2 Emissions in the Deep Ocean. Environm. Progress v.10, No1, Feb. 1991, pp. 64-74.

Hochstein, D.P. (1940) Carbon-Dioxide Power Cycle, Soviet Boiler and Turbine Construction, No10, pp. 420-423. (In Russian).

Hochstein, D.P., Kirov V.S. ed (1983), Cycles and Flowsheets of Nuclear Power Plants. Kiev, Vischaja schola (in Russian).

Holdren, J.P., Global Environmental Issues Related to Energy Supply. Energy v. 12 N 10/11, 1987, pp. 975-992.

Holt, T., Lindeberg, E. Thermal power without greenhouses gases and with improved oil recovery. p. 595.

Horlock, J.H., 1987, Cogeneration: Combined Heat and Power, Oxford, Pergamon.

Kabakov, V.I., Yantovskii, E.I. (1993) Solar Pond Magnetohydrodynamic Generator for Hydrogen Production Int. J. Hydrogen Energy, Vol. 18, No. 1, pp. 19-23.

Kantrovitz, A., Sporn, P. *Magnetohydrodynamics - Future Power Process*, Power Magazine, November 1959.

Kellog, H.K., Sizing up the Energy Requirements for Producing Primary Materials, Eng. and Mining J. 1977, v.178, No4, p. 61-65.

Krause, F., Bach, W., Koomey, J. Energy policy in the Greenhouse, London, Earthscan Publications, Ltd., 1990.

Kreith, F., West, R.E., Cleveland, C.J., Energy Analysis for Renewable Energy Technologies, ASHRAE Transaction, 1987, v.93, No1, p. 999-1010.

Kreuzer, H.J., *Nonequilibrium Thermodynamics*, Clarendon Press, Oxford (1984).

Laurmann, J.A. (1991) Energy Policy and Greenhouse Gas Induced Climatic Change, IJEEE v.1 No.1, pp. 51-61.

Linch, F.E., Egan, G.J., An Introduction Strategy for Hythane, an Alternative Fuel Blend of Hydrogen and Natural Gas. Proc. 2nd Annual CMDC Conf. Zurich, Dec. 1990, pp. 267-276.

Lodhi, M.A.K., Hydrogen Production from Renewable Sources, Int. J. Hydrogen En. v. 12, N7, 1987, pp. 461-468.

Lorentzen, G., Pettersen, J. (1990) Power Process Development for a Northern Climate. Eurogas '90. Proc from the Conference on Natural Gas. Trondheim, May 28-30, 1990, pp. 451-462.

Lozano, M.A., Valero, A., "Methodology for Calculating Exergy in Chemical Processes," 109 WAM of the ASME, Nov. 27-Dec.3, 1988, AES-v.4.

Lund, P.D., Kangas, M.T., Net Energy Analysis of District Solar Heating With Seasonal Heat Storage, Energy, 1983, v.8, No10, pp. 813-819.

Nordstrom, D.K., Munoz, J.L., Geochemical Thermodynamics. Blackwell Scientific Publications, 1986.

Oumov, N. The engineering tasks in view of dwindling of energy sources on the planet. Proc., V. 3, Moscow, Nature explorers Soc. 1916 (in Russian).

Parton, J.E., Owen, S.J.T., Raven, M.S., Applied Electromagnetics, McMillan London, 1986.

Pechtl, P. (1991) CO2 Emissionsminderung. Erdol und Kohle Petrochemie H.4, Apr., pp. 159-162.

Perelman, A.I. (1983) Geochemistry of Ground Waters. Nauka (in Russian)

Podrazhansky, A., Sagalevich, A., Zonnenshain, L.A.O., *Mysteries of the Deeps*, Moscow, Progress Publishers, 1989.

Posnansky, M., Technical and Economical Aspects of Solar Desalinations Desalination, v.67, 1987, pp. 81-95.

Rivkin S.L., Alexandrov A.A. Thermodynamic Properties of Water and Steam. Moscow, Energoatomizdat, 1984 (in Russian).

Schaefer, H.D., Die Nutzung Regenerativer Energiequellen in der Bundesrepublic Deutschland/ Energiewirtshaftliche Tagesfragen. 1987 v.3, No37, pp. 200-211.

Schilling, H.D., Bon, B., Krauss, U., *Kohlenvergasung*, Verlag Glükauf GMBH, Essen, 1981.

Seifritz W. (1989), Methanol as the Energy Vector. Int. Journ. Hydrogen Energy, v.14, No10, pp. 717-726.

Semenov, N.N. (editor), Solar Energy Conversion, Academy of Sciences of the USSR, Chernogolovka, in Russian, 1981.

Silvester, R.S., Enhanced oil recover-a literature review. Developments in oil recovery BHRA series No. 3, 1985.

Slesser, M., A Thermodynamic Constraint on the Rate of Global Development, Int. Chem. Eng. Symposium, v.1, Series No78, p. 37-48.

Slesser, M., Can Solar Energy Replace Fossil-Fissile Energy Sources, Solar Energy. 1980, v.25, p. 425-428.

Sokolov, E. Ya., Brodianskii, V.M., *Energy Background of Heat Transformation and Cooling*, M. Energoisdat, in Russian, 1983.

Steinberg, M., Albanese, A.S., Environmental Control Technology for Atmospheric Carbon Dioxie. In: Interaction of Energy and Climate, Proc. Int. Workshop held in Münster, Germany, March 3-6, 1980, D. Reidel Publisher, 1980, p. 521-551.

Stepanov, V.S., Energy and Exergy of Chemical Compounds Nauka, Novosibirsk, 1985 (in Russian).

Stewart, P.B., Munjal, P.K., *Journal of Chemical and Engineering Data*, 15(1)pp. 67-71, 1970.

Szargut, J., Morris, D., Cumulative Exergy Consumption and Cumulative Degree of Perfection of Chemical Processes, Energy Research, 1987, v.11, p. 245-261.

Szargut, J., Morris, D., Steward, F., Exergy Analysis of Thermal, Chemical and Metallurgical processes, NY, 1988.

Tabor, H., Israel Pioneers Solar Ponds Mod. Pow. Syst. July 1987, pp. 66-69.

Tsatsaronis, G., "A Course in Thermoeconomics," 111 WAM of the ASME, Dallas, Nov. 25-30, 1990.

Tsoulfanidis, N., Energy Analysis of Coal, Fission and Fusion Power Plants, Nuclear Technology, Fusion, 1981, v.1, p. 238-254.

van Engelenburg, B., K. Blok (1991), Prospect for the Disposal of Carbon Dioxide in Aquifers Vakgroep Natuur Wetenschap en Samenleving RU-Utrecht, G-91006, Utrecht, 14 Febr.

van Gool, W., Bruggink, J.J.C. (ed), Energy and Time in the Economic and Physical Sciences, Amsterdam, North Holland, 1985.

van Gool W. Hoogendoorn, Basic Aspects of Exergy Analysis and Exergy Optimization, "A future for Energy." Flowers' 90, p. 421-432.

van Steenderen, P. Carbon dioxide recovery from coal and natural gas fired combined cycle power plants by combustion in pure oxygen and recycled carbon dioxide. COMPRIMO Consulting Services BV, 03.03.1992.

Wall, G., Exergy-a Useful Concept, Phys. Res. Theory Group, Chalmers University of Technology, 1986.

Winter, C.J. Hydrogen Energy-Expected Engineering Breakthrough, Hydrogen energy, 1987, N.8, pp. 521-546.

Winter, C.J., Wasserstoff aus Sonnenenergie: ein additiver Energietrager, VGB Kraftwerkstechnic, 1989 v.69, H.3, pp. 245-255.

Yantovskii, E.I., *Gasdynamic Generator AC,* auth. sert. of March 31, 1959, no. 128542, Bulletin of Inventions, No8, 1960

Yantovskii, E.I., Poustovalov, Ju.V., Compressional Heat Pumps, Moscow Energoisdat, 1982 (in Russian).

Yantovskii, E.I., *Working Substances for Magnetogasdynamic Electrical Machines,* auth. sert. of September 1959, No. 128950, Bulletin of Inventions, No.11, 1960.

Yantovskii, E.I., Tolmach, E.M., *Magnetohydrodynamic Generators,* in Russian, M. Nauka, 1972.

Yantovskii, E.I., *Disk shaped MHD Generator of Hall Currents,* auth. sert. of June 4, 1965, No 187177, Bulletin of Inventions, No 2, 1966.

Yantovskii, E.I., *Non-equilibrium Thermodynamics in Thermal Engineering,* Energy v. 14, No7, pp. 393-390, 1989.

Yantovskii, E.I., Degtiarev, V.L. Internal Combustion Carbon Dioxide Power Cycles Without Exhaust Gases. Proc. Int. Conf. ENERGY SYSTEMS AND ECOLOGY, Cracow, Poland, 5-9 July, 1993, V. 1

Yantovskii, E.I., Zvagolsky, K.N., Gavrilenko, V.A. Computer Exergonomics of Power Plants Without Exhaust Gases. Energy Convers. Mgmt., V. 33, No. 5-8, pp. 405-412, 1992.

Yantovskii, E.I., Energy Supply Complex as a Condustor of Exergy Proc. of Intern. Conf. on Efficiency, Costs, Optimization and Simulation of Energy Systems, Zaragosa, Spain, June 15-18, 1992.

Yantovskii, E.I., Coal-fired MHD Generator Without Exhaust Gases. II Internat. MHD Conversion Conference, Beijing 1992, V. 1.

Zaitsev, I.K. Hydrogeochemistry of the USSR, L. Nedra, 1986, (in Russian).

Zaritski, S.P. Diagnostics of gas pumping units with gas turbine drive, Moscow, Nedra, 1987 (in Russian).

Ziegler, D., Lercher, G., Leeuwen, J. Pegus 12, The World Most Efficient Power Station. ASME 90-JPGC/GT-3, Boston, MA, Oct. 21-25, 1990.

Zurek, W.H., "Thermodynamic Cost of Computation, Algorithmic Complexity and the Information Metric," Nature, v.341, 14 Sept., 1989, pp. 119-124.

Zysin, V.A., *Combined Gas-Steam Installations and Cycles,* Gosenergoisdat, in Russian, Moscow, USSR, 1962.

Subject Index